T0310141

Wireless Automation as an Enabler for the Next Industrial Revolution

Wireless Automation as an Enabler for the Next Industrial Revolution

Edited by

Muhammad A. Imran, Sajjad Hussain and
Qammer H. Abbasi
James Watt School of Engineering
University of Glasgow
UK

This edition first published 2020

Registered Offices
John Wiley & Sons, Inc., 111 River Street, Hoboken, NJ 07030, USA
John Wiley & Sons Ltd, The Atrium, Southern Gate, Chichester, West Sussex, PO19 8SQ, UK

Editorial Office
The Atrium, Southern Gate, Chichester, West Sussex, PO19 8SQ, UK

For details of our global editorial offices, customer services, and more information about
Wiley products visit us at www.wiley.com.

Library of Congress Cataloging-in-Publication data applied for

HB ISBN: 9781119552611

Cover Design: Wiley
Cover Image: © Guryanov Andrey/Shutterstock

Set in 9.5/12.5pt STIXTwoText by SPi Global, Pondicherry, India

Printed and bound by CPI Group (UK) Ltd, Croydon, CR0 4YY

10 9 8 7 6 5 4 3 2 1

Contents

List of Contributors

Hasan T. Abbas
James Watt School of Engineering
University of Glasgow
UK

Muhammad Mahtab Alam
Thomas Johann Seebeck
Department of Electronics
Tallinn University of Technology
Estonia

Akram Alomainy
School of Electronic Engineering
and Computer Science
Queen Mary University of
London
UK

Ramy Amer
CONNECT Centre for Future
Networks
Trinity College Dublin
Ireland

Anas Amjad
Staffordshire University
UK

Gennaro Boggia
Department of Electrical and
Information Engineering
Politecnico di Bari
Italy

M. Majid Butt
Nokia Bell Labs
France
and
CONNECT Centre for Future
Networks
Trinity College Dublin
Ireland

Luigi Alfredo Grieco
Department of Electrical and
Information Engineering
Politecnico di Bari
Italy

Mona Jaber
School of Electronic Engineering
and Computer Science
Queen Mary University of
London
UK

Alar Kuusik
Thomas Johann Seebeck
Department of Electronics
Tallinn University of Technology
Estonia

Yannick Le Moullec
Thomas Johann Seebeck
Department of Electronics
Tallinn University of Technology
Estonia

Hassan Malik
Thomas Johann Seebeck
Department of Electronics
Tallinn University of Technology
Estonia

Nicola Marchetti
CONNECT Centre for Future
Networks
Trinity College Dublin
Ireland

Zhen Meng
James Watt School of Engineering
University of Glasgow
UK

Sven Pärand
Telia Estonia Ltd.
Estonia

João Pedro Battistella Nadas
James Watt School of Engineering
University of Glasgow
UK

Metin Ozturk
James Watt School of Engineering
University of Glasgow
UK

Mohsin Raza
Faculty of Science and
Technology
Middlesex University
UK

Aifeng Ren
James Watt School of Engineering
University of Glasgow
UK

Huan X. Nguyen
Faculty of Science and
Technology
Middlesex University
UK

Hafiz Husnain Raza Sherazi
Department of Electrical and
Information Engineering
Politecnico di Bari
Italy

Richard Demo Souza
Universidade Federal de Santa
Catarina-Florianópolis
Brazil

Adnan Zahid
James Watt School of Engineering
University of Glasgow
UK

Guodong Zhao
James Watt School of Engineering
University of Glasgow
UK

Ahmed Zoha
James Watt School of Engineering
University of Glasgow
UK

Preface

Over the past three centuries, industrial processes have evolved from steam operated mechanical looms (industry 1.0), electrically run machines (industry 2.0) to programmable logic controller (PLC) based manufacturing (industry 3.0). However, in the 21st century, we are talking about the industrial revolution through to industry 4.0. Industry 4.0 is about digitizing all industrial processes and making them intelligent, efficient and autonomous through sensing, connectivity, big data analytics, and control. With advancements in enabling technologies, industrial processes are likely to build further on industry 4.0, resulting in even smarter manufacturing solutions.

Besides the technical challenges, the industrial revolution will face challenges like overhauling of company culture, and hiring and training people with appropriate skills to run the digital processes. Moreover, the companies will need to develop new business models in order to maximize their outputs and take full advantage of the industrial revolution.

With regards to the technical challenges, cybersecurity is an extremely important issue since any sort of cyber attack could result in financial as well as human loss. Interoperability is another important aspect that has to be ensured in the form of the use of open source software and solutions, privacy and accessibility.

The key enablers for industrial revolution are advanced electronics and information and communication technologies (ICT), e.g. energy-efficient sensing, cloud and mobile-edge computing, reliable and low latency communication, big data analytics through machine learning and artificial intelligence, and the Industrial Internet of Things (IIoT).

To enable flexible and scalable connectivity solutions for industry 4.0, the role of wireless automation is pivotal. To achieve industrial wireless automation there are stringent requirements of low latency and high reliability in the presence of harsh industrial environments. Specifically, to enable industrial revolution, the end-to-end latency should be less than 0.5 ms while the reliability requirements of 10–9 or even less block error ratio (BLER).

To address these issues, the fifth generation of mobile communications (5G) proposes meeting the industrial wireless automation requirements through ultra-reliable low-latency communication (URLLC) services.

Integrating 5G URLLC with machine learning and artificial intelligence solutions will result in proactive anomaly detection followed up by self-healing enabled by a reliable feedback control loop.

This book is presented in a way that it gradually builds on the knowledge introduced in the previous chapters, thus providing a seamless integration of various topics. The book chapters can be broadly divided into the following parts:

- Industrial wireless sensor networks, application, challenges, and solutions (Chapters 1–3)
  - Industrial wireless sensor networks and their applications
  - Life-span extension for sensor networks in the industry
  - Multiple access and resource sharing for low latency critical industrial networks
- Industrial automation via 5G URLLC and IIoT (Chapters 4 and 5)
  - IIoTs and narrow-band IoT for industrial automation
  - URLLC as an enabler for industry automation
- Machine learning and industrial automation optimization (Chapters 6–8)
  - Anomaly detection and self-healing in industrial wireless networks
  - Cost efficiency optimization for industrial automation
  - Non-intrusive load monitoring

- Advanced topics on wireless network control, nano-scale communication and wireless caching (Chapters 9–11)
  - Wireless networked control
  - Wireless caching for industrial automation
  - Nano-scale communication for agriculture industrial automation.

Muhammad A. Imran, Sajjad Hussain,
Qammer H. Abbasi
University of Glasgow, UK

1

Industrial Wireless Sensor Networks Overview

Mohsin Raza and Huan X. Nguyen

Faculty of Science and Technology, Middlesex University, UK

1.1 Introduction

Over the past few years, industrial evolution has led to the development of more sophisticated and effective supporting technologies to improve performance in a wide range of industries. Industrial wireless sensor networks (IWSNs), as a suitable alternate to the wired links, have the potential to play an important role in optimizing data communications within industries. Various attributes of IWSNs, including low installation costs, self-healing abilities, flexibility, self-organization, localized processing, interoperability, scalability, and ease of deployment, have been very influential in wide-scale adoption of the technology. Furthermore, IWSNs also offer the potential to operate in critical and emergency applications as well as non-critical monitoring applications (Raza et al., 2018a).

This chapter provides an overview of the IWSNs. A detailed discussion is provided on industrial systems and the requirements of IWSNs. In addition, applications of IWSNs and research developments are also discussed.

Demand for high productivity to cope with the rising needs has resulted in massive expansion of industries. The advancements in manufacturing processes and the introduction of a high level of automation has resulted in improved productivity and process efficiency. The increasing number of products and customer demands

Wireless Automation as an Enabler for the Next Industrial Revolution, First Edition.
Edited by Muhammad A. Imran, Sajjad Hussain and Qammer H. Abbasi.

have encouraged researchers to uncover the possibility of increasing the overall production efficiency. The recent developments in micro-electromechanical systems (MEMS), radio communications and control systems have been very influential in pushing the limits of technology (Raza et al., 2018b). These developments have brought a new era of manufacturing and automation within the industries.

The high-speed developments within the industrial sector are highly influential in defining the standard operating procedures and practices. The history of industrial developments can be described in terms of industrial revolution (Raza et al., 2018a).

Industrial revolution begins with many technological innovations in manufacturing and processing. The first industrial revolution was mainly powered by coal and steam to run the machines. In the second revolution the focus shifted to mass production models with the higher degree of inclusion of machines operated by electric power. The third revolution incorporated computer driven systems and introduced certain degrees of automation to the industrial processes. The introduction of digital technologies has paved the path for future developments in industry. These gradual improvements set the path for a fully automated self-reliant industry in which machine intelligence, communications and distributed control systems enabled the fourth industrial revolution (Industry 4.0) (Yahya, 2017). A brief overview of the four industrial revolutions and their salient features is presented in Table 1.1.

Table 1.1 Salient features of the industrial revolutions.

No.	Industrial revolution	Salient features
1.	First industrial revolution	Coal and steam powered technology, mechanical innovations
2.	Second industrial revolution	Electric powered mass production models
3.	Third industrial revolution	Use of digital technologies, utilization of greener energy sources, introduction of partially automated processes
4.	Fourth industrial revolution	Fully automated self-reliant systems, interconnected intelligent systems

Industry 4.0 offers a conceptual platform for further developments for industry and other sectors that incorporate a smart interconnected pervasive environment. Industry 4.0 benefits from diverse technologies including a digital supply chain, mass production, ad hoc communication networks, sensors web, big data platforms, distributed control systems, artificial intelligence and machine learning, sensors and actor networks, next-generation robotics and digital twins. All these technologies provide opportunities for future growth and innovation in industry 4.0. The developments in industry 4.0 focus on integration of processes through an extensive communication networks to support the development of smart processes and products. Along with the incorporation of self-sustaining processes, industry 4.0 offers optimized effective and efficient processes to improve production time and efficiency (Garrido-Hidalgo et al., 2018).

Effective communication among different processes within industries offer robust, energy efficient and timely operation. It also allows identification and elimination of any bottlenecks in different disjoint processes (Raza et al., 2017). In addition, effective communication also offers a means for smart self-sustained systems with the incorporation of classical and deep machine learning techniques. Interlinking the sensing and decision systems allows large scale proactive integration among the processes, thus forming an intelligent industrial environment to facilitate smart decisions.

1.2   Industry 4.0

Industry 4.0 depicts the current trend and vision for future transformations. The main driving force for industry 4.0 is the innovation in automation, robotics, machine intelligence, data exchange and smart manufacturing. Industry 4.0 is not a stand-alone system, rather it incorporates the Internet of Things (IoT), sensor networks, fog and edge computing, distributed control systems, digital twins and cyber physical systems.

Industry 4.0 empowered by cyber physical systems aims at establishing smart factories where all the processes and actions are not only actively monitored but also interlinked for improved system efficiency (Yahya, 2017). It incorporates a modular approach to enhance flexibility and scalability where cyber physical systems and digital twins provide computer based analysis and control of the critical processes. The cyber physical systems and digital twins have notable significance in industry 4.0, as they can assist in replication of virtual processes and virtualization of the functional systems to help improve the overall process and to make decentralized decisions. Wireless sensor networks (WSNs) enable the cyber physical systems to predict possible alterations in the industrial processes to achieve higher flexibility. Collectively, these systems establish a real-time self-cooperative environment to support critical decisions and manage operational functionality across organizational services (Qi and Tao , 2018).

Industry 4.0 is driven by interconnection, information transparency, assistive systems and decentralized decisions. Interconnection interlinks sensors, actuators, control centres, devices and people, allowing them to make smart decisions. Interconnectivity also help in making smart decisions in a modular framework.

Information transparency offers access to extensive data generated by the plant. This allows the operators to make suitable decisions where needed. The information transparency also allows the effective use of a vast amount of useful information to make environment friendly decisions and to reduce the carbon footprint. It also opens up avenues for transfer learning.

Assistive systems serve as a backbone for further developments in industry 4.0. The improvements in this domain will highly depend on the ability of smart assistive systems to facilitate decision making in highly complex environments. Assistive systems are expected to provide humans with the graphical data analytics to make informed decisions, even in complex processes where several hundreds of thousands of entries are recorded on a daily and weekly basis. In addition, assistive systems will also be able to assist humans in performing the exhaustive and physically unsafe operations using advanced robotics.

Decentralized decisions improve the ability of industry 4.0 to make timely decisions by cutting down the overall round trip time to and from the control centre (Raza et al., 2018a). These decisions help in improving the system efficiency and also allow to meet the stringent time deadlines in emergency and regulatory systems. In addition, these decentralized decisions also allow the industries

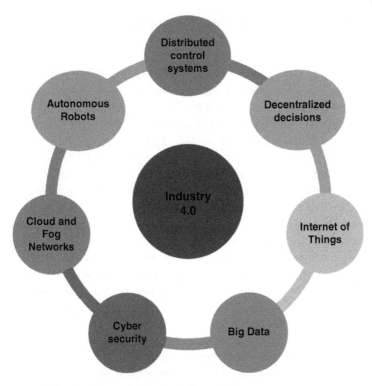

Figure 1.1 Enabling technologies, industry 4.0.

to work autonomously without the intervention of humans. Thus, human intervention and feedback are only required in extremely critical and conflicting decisions (Raza et al., 2018b). The flow diagram of key enabling technologies in industry 4.0 is presented in Figure 1.1.

Industry 4.0 offers notable potential for envisioned developments; however, there are several limitations that require further investigation to overcome challenges in this domain. One such challenge is the communication infrastructure to enable timely and reliable communications within the industrial processes, devices, and control centres. IWSNs are one such domain with the potential to overcome these limitations. The upcoming passages in this chapter discusses the details of different attributes of IWSNs and possible research potential for future developments.

1.3 Industrial Wireless Sensor Networks (IWSNs)

WSNs provide extensive coverage with spatially distributed sensor motes. The low-cost multi-hop communications in WSNs offer the necessary technology to support diverse applications. In WSNs, wireless motes are equipped with sensors, a battery, radio, processing capabilities and memory (Raza et al., 2011; Raza, 2018). A graphical representation of sensor nodes is presented in Figure 1.2.

The wireless motes in WSNs can be broadly divided into sensor, relay and sink nodes. Sensor nodes sample the immediate environment. Once the sensor value is read, it is transmitted through radio to be delivered to the sink. Thus, the sink node serves as a data collection centre where all the sensor readings are collected. The sink sometimes serves as a gateway between the WSN and the control centre. Since the wireless nodes in WSNs use short range communications, information is delivered from sensor node to sink in a multi-hop fashion. The relay nodes facilitate the multi-hop communications by forwarding received packets to the relevant node towards the destination until it reaches the sink (Kumar et al., 2011). A graphical representation of a typical WSN is presented in Figure 1.3.

WSNs are widely used for monitoring applications. However, their applications have recently emerged in industry. Since the requirements of traditional WSN applications and applications within industry can vary notably, such WSNs are referred as IWSNs. IWSNs deal with applications from six domains of industry: emergency systems,

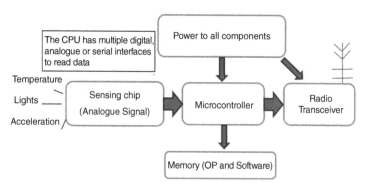

Figure 1.2 Block diagram of a typical wireless sensor mote.

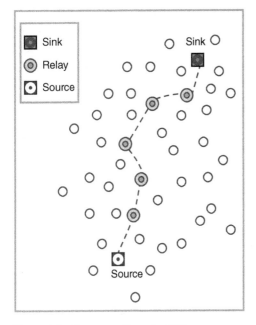

Figure 1.3 Representation of a WSN.

regulatory control systems, supervisory systems, open loop control systems, alerting systems and monitoring systems (Raza et al., 2018b).

Emergency systems deal with the critical processes within the industrial setup and are highly critical for safety in industrial plants. In such systems communication plays an important role and timely communication of any critical scenario is very important (Somappa et al., 2014).

Regulatory control systems deal with regulated processes, where a continuous feedback from the sensors to the control unit is required for stabilized processes. In regulatory control systems, ultra-reliable communications are required to keep the processes running with full efficiency. The process stabilization can be affected if the information packets are delayed or corrupted (Raza, 2018).

Supervisory control systems deal with relatively less critical processes that only require data communications from sensor nodes if certain sensory thresholds are violated. The communications in supervisory control systems are asymmetric in nature and are less critical compared to regulatory control systems (Raza et al., 2018b).

Open-loop control systems usually deal with the processes with relatively high delay tolerance. The information in open-loop control systems is usually communicated to the control centre periodically (every few seconds to minutes) to see the overall process is working effectively. The information received in open loop control systems is not directly used as feedback for process control, rather, it is observed by operators, to take suitable actions if needed.

Alerting systems deal with low-priority processes that occasionally communicate with the control unit. The main purpose of alerting systems is to highlight the undesirable reading received from different processes. Since these systems do not deal with critical processes, no critical impact on the processes can be expected in the event of delay in communications of such data (Raza et al., 2018b).

Monitoring systems implement data gathering applications where there is no viable stake for missing any communications or delay in receiving the readings. The communication of monitoring data is dependent on the availability of communication resources and accumulated data is mainly used to plan future developments.

To deal with the critical applications in emergency, regulatory control and supervisory systems, more critical and stringent requirements need to be met by IWSNs. Therefore, IWSNs need to communicate more reliably and within strict time deadlines.

Research and development, and continuous improvements in IWSNs have resulted in their widespread usage across different industries. The advantages of cost efficiency, scalability, self-healing, flexibility, ease of deployment and reformation also encourage the wider acceptability of IWSNs. Some of the applications of IWSNs are listed as follows.

1.4 Applications of IWSNs

The applications of WSNs vary from sophisticated feedback control, emergency and safety to monitoring and data accumulation. These applications are discussed in the upcoming sections.

1.4.1 Feedback Control Systems

One of the emerging applications of IWSNs is feedback control systems where periodic readings are desired from the sensor nodes to maintain certain operational stability and efficiency. Applications in feedback control system are distributed to manage diverse industrial requirements, where some of these include vibration stabilization, pressure regulation, maintenance and detection.

1.4.2 Motion and Robotics

One of the most common applications of IWSNs in emerging industry 4.0 is incorporation and control of motion and robotics. The high degree of movement freedom in robotics requires a high level of automation. However, due to the increased movement and relatively short window of response in robotics, the time deadlines for communications are relatively less.

1.4.3 Safety Applications

In industry 4.0, several applications deal with a time sensitive and high risk environment (Raza et al., 2018b). In such applications, ultra-reliable and low latency communications are required from IWSNs. One such application is the mining industry, where the possibility of the accumulation of toxic and flammable gases cannot be denied. For such circumstances, the IWSNs are required to identify such threats in a timely way, and warn workers and vulnerable employees. Use of IWSNs in safety applications allows the smooth running of processes and countermeasures to negate any potential hazard, which can become threats if left undetected.

1.4.4 Environmental Monitoring

Environmental monitoring is a term widely used to monitor the impact of industries on the natural habitat. Such monitoring networks accumulate information regarding pollutants released in water and air. The accumulated information regarding industrial waste can be used to effectively improve the processes to reduce the carbon footprint of a particular industry. IWSNs can play a vital role in monitoring air quality, water pollution, soil contamination, etc. In addition to monitoring, IWSNs can also play a role in evaluating the real-time proportion of gases or water contamination, thus giving an extensive analysis of waste discharge of manufacturing plants and industrial installations.

1.4.5 Machine/Structural Health Monitoring

The continuous vibration generated from operating machinery within industry and aging factors weaken the infrastructure. In addition, the machinery also requires regular maintenance. IWSNs have been used to evaluate the machinery condition and/or structural health. The sensor nodes are deployed to evaluate conditions of various potentially important machinery parts and to evaluate the need for machinery maintenance. Therefore, suitable use of wireless nodes is widely adopted to access information related to machine/structural health.

1.5 Communication Topologies in IWSNs

The communication within IWSNs usually takes place in one of the three main topologies: star, mesh and tree (Raza et al., 2018b).

In star topology, the sensor nodes are directly connected to the cluster head (AlShawi et al., 2012; Yang et al., 2010). Due to the direct connection to the cluster head/decisioning unit, the overall communications delay is minimised. Time division multiple access (TDMA) is mostly used in communications scheduled in star topologies. TDMA allows the devices to establish interference free communications. However, in some cases carrier sense multiple access collision avoidance (CSMA/CA) is also used. A graphical representation of the star topology is presented in Figure 1.4.

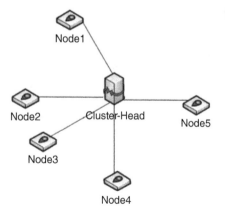

Figure 1.4 Star topology.

Figure 1.5 Mesh topology.

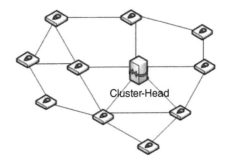

Mesh topology is used for larger networks. Mesh topology provides better connectivity between the nodes and thus improves overall reliability and connectivity in the case of larger networks (Raza et al., 2018b). However, it also provides added delay because of allowing multiple links to the gateway and the flexibility to opt for the most stable route for information communication (Riggio et al., 2011). This topology connects multiple nodes to every node, allowing improved reliability along with self-healing abilities (Vinh et al., 2012). A graphical representation of the mesh topology is presented in Figure 1.5.

Tree topology provides dedicated links to minimize the overall communication overhead. The communications in tree topology are sorted before initiating the communication and usually a node in the network will take a fixed number of hops to reach the destination, unless the communication path is revised, which adds deterministic behavior to the communication. Tree topology uses gradient information to route data to the sink (Wu et al., 2008). Since the information communications in tree topology is multi-hop, failure in a link or key linking nodes can affect large network branches. A graphical representation of the tree topology is presented in Figure 1.6.

1.6 Research Developments and Communications Standards for Industry

Several communication standards and industrial protocols are introduced to facilitate communications within industries. These standards primarily target the flow of information from sensors to the control centre using the industrial, scientific and medical (ISM)

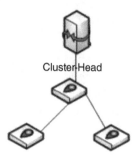

Figure 1.6 Tree topology.

Cluster Head

radio band. IEEE WPAN standards operate in the ISM band to facilitate wireless communications. The IEEE standards (IEEE802.15.4, IEEE802.15.4e) provide a framework to support further developments in the communication infrastructure to enable applicability of WSNs within the household, buildings, healthcare and industry.

Apart from the IEEE 802.15.4 (IEEE Standard for Information Technology, 2006) and 802.15.4e (IEEE Standard for Local and Metropolitan Area Networks, 2012) standards, which offer physical and MAC layer specifications, certain industrial protocols are also introduced that provide upper layer specifications. Some of the well known standards include ZigBee, WirelessHART, ISA100.11a, and 6LoWPAN (IETF Datatracker, 2018; Lee et al., 2007). Further details on IEEE standards and industrial protocols are as follows.

1.6.1 IEEE 802.15.4

IEEE 802.15.4 specifies the physical layer and MAC layer in WSNs. The specification IEEE 802.15.4 targets low power communications with modest data rates to support energy efficient communications between nearby devices.

The basic specifications target low power self-sustainable and self-healing networks with the ability of autonomous operation. The standard lays out the functional overview, frame structure, security and management services, transmission power control, modulation, etc.

It also specifies the frequency band, data parameters and spreading parameters. The MAC layer specifications in IEEE 802.15.4 handles beacon generation, synchronization, medium access assurance, security, association and disassociation. It also specifies the roles of different devices within the network to distribute responsibilities.

1.6.2   IEEE 802.15.4e

IEEE 802.15.4e implements the same physical layer parameters as IEEE 802.15.4. However, notable changes have been introduced in the MAC layer to support communications within the industrial environments and industrial systems. The channel access is changed from CSMA/CA to TDMA to better facilitate the communications for regulatory and supervisory control systems. All the amendments in the existing IEEE 802.15.4 standard are focused on enhancing suitability for existing critical industrial applications.

TDMA based channel access also allows guaranteed channel access and congestion minimization in contrast to CSMA/CA based channel access. The use of synchronization beacon is also introduced to synchronize the communications from multiple sensor nodes in a single frame. In addition to regular TDMA based distribution of slots, shared slots are also introduced to allow contention based channel access in case retransmissions are required.

1.6.3   Zigbee

ZigBee provides the architecture for additional layers in the IEEE 802.15.4 standard. Zigbee uses the physical and MAC layer specification of the IEEE standard and adds a network layer, an application sub-support layer and an upper application layer. The data rate is relatively modest (250 kbps) but suitable for daily life monitoring and feedback applications (Utility Industry, 2018). The frequency channels in Zigbee are divided into 16 2 MHz channels. In addition, the extended addressing mechanism allows connection with 2^{16} devices at once. Zigbee allows the formation of mesh, tree and star topologies, which further extends its applicability to wider application areas. The open source nature of Zigbee has encouraged its wider acceptability and currently over 70 million ZigBee devices are operational worldwide (Utility Industry, 2018).

While Zigbee offers many benefits, there are certain aspects that limit its applications in time critical and reliability centric applications. Since Zigbee uses a CSMA/CA based channel access scheme, the channel access cannot be guaranteed proactively. Rather, it follows the contention based channel access. In addition, Zigbee shares some overlapping channels with Wi-Fi that may cause interference where both the technologies are operating simultaneously (Raza et al., 2018b).

1.6.4 WirelessHART

WirelessHART is developed by HART communications to support IWSNs. It primarily targets applications in regulatory control, supervisory control, open loop control, alerting and monitoring systems. Estimates suggest over 30 million WirelessHART devices in operation (WirelessHART, 2016). Like Zigbee, WirelessHART can also establish large networks and can support different communication topologies. The protocol uses time slotted access and also benefits from the specifications of IEEE 802.15.4e to enable extended industrial applications. Handshaking, communications and affiliation are validated with private key issuance. In addition, channel hopping is incorporated to enhance security and safekeeping against eavesdropping.

While WirelessHART offers several features that complement its suitability in industry, it fails to offer appropriate solutions to facilitate interoperability. It is also not compatible with IP based devices and the IoT.

1.6.5 ISA100.11a

ISA100.11a is a wireless network solution for IWSNs. It targets industrial applications in automation, process control and monitoring. ISA100.11a also uses a similar infrastructure to WirelessHART. Apart from the physical layer parameter similarity, TDMA based access and channel hopping for security are also implemented. ISA100.11a also implements extended MAC for added features. In addition, the specification of an upper data link layer, network layer, UDP and TCP and application layer are also defined. ISA100.11a is also IP enabled and supports IPv6. The network and transport layer in ISA100.11a allows the optional implementation of a CSMA/CA based exponential back-off mechanism.

1.6.6 6LoWPAN

6LoWPAN is a IPV6 based low-power wireless personal area network (Lee et al., 2007). It can be used to directly connect with IP devices and wired solutions like real-time ethernet (commonly used wired solution for data communications within industry). Suitable security features are also implemented with IPv6 adaptation. The use of AES-128 is incorporated for enhanced security. Since the protocol is developed to offer low power consumption, inactive and active regions are designated within the superframe to allow extended sleep cycles for the nodes and coordinators. In addition to the active and inactive regions, the protocol also introduces low-power listening modes for the nodes to enhance energy conservation. The network formation in 6LoWPAN can be based on IP as well as MAC addresses to interconnect non-IP devices with IP-enabled devices. While 6LoWPAN offers many benefits, from the use of IPv6 to extended security, it also suffers from certain limitations posed by CSMA/CA based channel access. Modest data rates from 20 to 250 kbps also add limitations in multimedia and periodic data communication applications.

Bibliography

AlShawi S., Yan L., Pan W. and Luo B. (2012) Lifetime enhancement in wireless sensor networks using fuzzy approach and A-star algorithm *IEEE Sensors J.*, 12, 10, 3010–3018.

Field Wireless (2018). ISA-100 Wireless Compliance Institute, ISA-100 Wireless, Compliance Institute-Official Site of ISA100 Wireless Standard. [Online]. Available: http://www.isa100wci.org/

Garrido-Hidalgo C., Hortelano D., Roda-Sanchez L., Olivares T., Ruiz M. C. and Lopez V. (2018) IoT Heterogeneous Mesh Network Deployment for Human-in-the-Loop Challenges Towards a Social and Sustainable Industry 4.0. *IEEE Access*, 6, 28417–28437. https:doi:10 .1109/ACCESS.2018.283667

IEEE (2006) IEEE Standard for Information Technology—Local and Metropolitan Area Networks—Specific Requirements—Part 15.4: Wireless Medium Access Control (MAC) and Physical Layer (PHY) Specifications for Low Rate Wireless Personal Area Networks (WPANs), Standard 802.15.4-2006, IEEE, London, 1–320.

IEEE (2012) IEEE Standard for Local and Metropolitan Area Networks—Part 15.4: Low-Rate Wireless Personal Area Networks (LR-WPANs) Amendment 1: MAC Sublayer, Standard 802.15.4e-2012, IEEE, London, 1–225.

IETF Datatracker (2018) 6loWPAN Active Drafts. [Online]. Available: https://datatracker.ietf.org/doc/search/?name=6loWPAN& activeDrafts=on&rfcs=on

Kumar D., Aseri T. C., and Patel R. B. (2011) Multi-hop communication routing (MCR) protocol for heterogeneous wireless sensor networks. *International Journal of Information Technology, Communications and Convergence* 1, 2, 130–145.

Lee J.-S., Su Y.-W. and Shen C.-C. (2007) A comparative study of wireless protocols: Bluetooth, UWB, ZigBee, and Wi-Fi. *Proc. 33rd Ann. Conf. IEEE Ind. Electron. Soc. (IECON), Taipei, Taiwan, 2007*, 46–51. https:doi:10.1109/IECON.2007.4460126

Lee J.-S., Su Y.-W. and Shen C.-C. (2007) A comparative study of wireless protocols: Bluetooth, UWB, ZigBee, and Wi-Fi. *Proc. 33rd Annu. Conf. IEEE Ind. Electron. Soc. (IECON), Taipei, Taiwan, 2007*, 46–51. https:doi:10.1109/IECON.2007.4460126

Qi Q. and Tao F. (2018) Digital Twin and Big Data Towards Smart Manufacturing and Industry 4.0: 360 Degree Comparison. *IEEE Access*, 6, 3585–3593. https:doi:10.1109/ACCESS.2018.2793265

Raza M., Ahmed G., Khan N. M., Awais M. and Badar Q. (2011) A comparative analysis of energy-aware routing protocols in wireless sensor networks *2011 International Conference on Information and Communication Technologies, Karachi, 2011*, 1–5. https:doi:10.1109/ICICT.2011.5983546

Raza M., Hussain S., Le-Minh H., Aslam N. (2017) Novel MAC layer proposal for Ultra-Reliable and Low-Latency Communication in Industrial Wireless Sensor Networks; A proposed scenario for 5G powered Networks. *ZTE Communications*, S1, 006, 1673–5188.

Raza M. et al. (2018) Dynamic Priority Based Reliable Real-Time Communications for Infrastructure-Less Networks *IEEE Access*, 6, 67338-67359, 2018. https:doi:10.1109/ACCESS.2018.2879186

Raza M., Aslam N., Le-Minh H., Hussain S., Cao Y. and Khan N. M., (2018a), A Critical analysis of research potential, challenges, and future directives in industrial wireless sensor networks. *IEEE Communications Surveys & Tutorials*, 20, 1, 39–95. https://doi:10.1109/COMST.2017.2759725

Raza M., Aslam N., Le-Minh H., Hussain S. (2018b) A Novel MAC proposal for Critical and Emergency Communication in Industrial Wireless Sensor Networks. *Elsevier Computers and Electrical Engineering*, 72, 976–989. https://doi.org/10.1016/j.compeleceng.2018.02.027.

Riggio R., Rasheed T., and Sicari S. (2011) Performance evaluation of an hybrid mesh and sensor network. *Proc. IEEE Int. Glob. Telecommun., Kathmandu, Nepal, 2011*, 1–6.

Shelby Z. and Bormann C. (2011) 6LoWPAN: The Wireless Embedded Internet, 43, Wiley, New York, NY.

Somappa A. A. K., Ovsthus K. and Kristensen L. M. (2014) An industrial perspective on wireless sensor networks—A survey of requirements, protocols, and challenges. *IEEE Communications Surveys & Tutorials*, 16, 3, 1391–1412.

Vinh T. T., Quynh T. N. and Quynh M. B. T. (2012) EMRP: Energy-aware mesh routing protocol for wireless sensor networks. *Proc. IEEE Int. Conf. Adv. Technol. Commun., Hanoi, Vietnam, 2012*, 78–82.

WirelessHART (2019).WirelessHART-flexibility in process management [Online]. Available: https://www.uk.endress.com/en/solutions-lowering-costs/field-network-engineering/fieldbus-technology/wirelesshart-communication-fieldbus-technology

Wu Y., Fahmy S. and Shroff N. B. (2008) On the construction of a maximum-lifetime data gathering tree in sensor networks: NPcompleteness and approximation algorithm. *Proc. INFOCOM, Phoenix, AZ, USA, 2008*, 131–136.

Yahya A. (2017) 4th Industrial revolution: The future of machining. *4th International Conference on Information Technology, Computer, and Electrical Engineering (ICITACEE), Semarang, 2017*, 3–4. https:doi:10 .1109/ICITACEE.2017.8257664

Yang F., He C., and Shao H.-Z. (2010) Two simplified coding schemes for wireless sensor networks with star topologies *Proc. Int. Conf. Softw. Telecommun. Comput. Netw. (SoftCOM)*, 185–189.

Zigbee Alliance (2018). Utility Industry, The ZigBee Alliance, Davis, CA. [Online]. Available: http://www.zigbee.org/what-is-zigbee/utility-industry/

2

Life-span Extension for Sensor Networks in the Industry

Metin Ozturk[1], Mona Jaber[2], and Muhammad A. Imran[1]

[1] *James Watt School of Engineering, University of Glasgow, UK*
[2] *School of Electronic Engineering and Computer Science, Queen Mary University of London, UK*

2.1 Introduction

The proliferation of sensor networks (SNs) has made them an integral part of various domains including industry, healthcare, and agriculture. SNs reduce the need for human intervention; hence they are able to make operations quicker and more cost effective Gungor und Hancke (2009).

Although different applications employ SNs for different purposes, their fundamental duty is to acquire physical parameters from an environment of interest, such as temperature and vibrations. In agriculture, for example, they can be used to monitor a farm in order to update a remote body, e.g. a farmer, in this case, can be updated with changes in humidity levels of a crop, in order to improve the production quality Jawad *et al.* (2017). As such, SNs enable informed, accurate, and remote decision making with the possibility of feeding back the desired response to the actuators Queiroz *et al.* (2017). This complete loop is often referred to as a cyber physical system in which the actuators are autonomously triggered without human interaction Erdelj *et al.* (2013). For example, if sensors are reporting lower than desired humidity level in the soil, the SN will report the measurements to the server and sprinklers could be remotely activated automatically or manually.

Wireless Automation as an Enabler for the Next Industrial Revolution, First Edition.
Edited by Muhammad A. Imran, Sajjad Hussain and Qammer H. Abbasi.
© 2020 John Wiley & Sons Ltd. Published 2020 by John Wiley & Sons Ltd.

In the case of healthcare, on the other hand, SNs can be used to collect data from human beings or animals to manage their health conditions by monitoring body-related parameters, e.g. blood pressure Du *et al.* (2011), or even tracking pills inside of a body Fernández *et al.* (2018).

SNs can also play a crucial role in industry scenarios with a wide range of applications including inventory management and industrial safety Gungor und Hancke (2009); Kolavennu and Gonia (2016). The rapidly developing industrial sector is causing an increase in investment commitments. Consequently, operational efficiency has become a crucial requirement. In addition, the competition among companies compels them to be agile and dynamic by adapting new technologies to increase (or at least sustain) their market share. Therefore, the main focus of this chapter is to discuss SNs from an industrial perspective.

We distinguish two categories of SNs based on the adopted communication method: wireless SNs (WSNs) and wired SNs. As noted by Hodge *et al.* (2015), wired connectivity requires cumbersome and costly installations and maintenance efforts; hence they are less attractive than WSNs. Particularly in the domain of industrial SNs, the usage of wired connectivity is more challenging in view of the harsh environment that would need additional consideration to protect the network from effects such as vibration, heat, and humidity Low *et al.* (2005).

WSNs, on the other hand, are more efficient and practical, not only because their installation is much easier and cheaper, but also because they require far less maintenance. In addition, WSNs can be relocated easily in the case of any environmental problem and/or rearrangement of industrial plants.

Today's competitive business environment entails being dynamic and agile, which is very hard without including some sort of intelligence in systems. To this end, WSNs are quite open to intelligent solutions, such as self-organization with its three main branches: self-configuration, self-optimization, and self-healing Aliu *et al.* (2013). Based on the presented comparison and the dominance of WSNs in the industrial vertical, the rest of the chapter focuses on WSNs as the prevailing SN solution. Nevertheless, some of the technologies described in this chapter apply equally to wired SNs and WSNs, such as life-span extension methods.

The life-span of WSNs is a crucial feature that is essential to delivering the anticipated advantages, such as cuts in operational expenditure and improved efficiency. However, it is expected that their life-span should be in the order of years Anastasi *et al.* (2009) to render the solution profitable. If, in contrast, WSN behaviour were to be repaired or replaced more often, the incurred cost would impede the operational gain. This aspect is particularly applicable to industrial WSNs, which are often deployed in a random fashion Anastasi *et al.* (2009).

WSN life-span is determined by their energy balance; WSNs are often powered by batteries owing to their convenience, albeit limited power, which adds a hard constraint to the energy balance. In other words, sensor nodes are energy constraint devices, and thus the life-span of WSNs can be extended only if their energy balance is increased. This chapter will focus on state-of-the-art life-span extension methods for WSNs with high-level taxonomy: *energy harvesting* and *energy conservation*.

The organization of this chapter is as follows: in Section 2.2, WSNs and their working principles are presented in detail. Section 2.3 discusses industrial WSNs with their characteristics and some selected applications, while Section 2.4 elaborates the life-span extension methods for WSNs, and introduces an artificial intelligence (AI) aided method of optimizing energy efficiency jointly with application-centric requirements. Lastly, Section 2.5 concludes the chapter.

2.2 Wireless Sensor Networks

As seen in Figure 2.1, a typical WSN consists of multiple sensor nodes, a sink node, and an end user. The sensor nodes collect information about physical parameters from an environment, such as temperature and humidity, and transmit it to the sink node, which acts as a base station using a wireless connection. According to the type of application, the collected data can be processed locally or at the remote edge, giving the sink node two options:

- it can send the data to a local end-user;
- it can connect to the internet to transmit the data to a remote end-user.

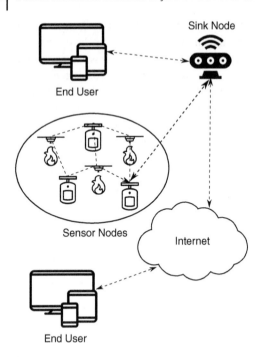

Figure 2.1 General structure of wireless sensor networks.

The latter utilizes the concept of the Internet of Things (IoT), where all devices have access to the internet for communication.

Figure 2.2 showcases the architecture for a typical sensor node with its main components: battery, sensing unit, micro-controller unit (μCU), memory, and radio unit. These components enable sensor nodes to execute three main tasks including sensing, processing, and communicating. Each of these tasks performed in a typical sensor node results in energy consumption, and thus a battery is greatly needed as a power supply. The downside of using batteries in the system is that they are typically limited energy sources, making the life-span of a node dependent on the amount of energy stored in its battery. However, batteries are still convenient solutions for WSNs, since in many cases they are deployed over wide areas in a random fashion Hodge *et al.* (2015) for inaccessible and dangerous regions, where pre-determination of sensor locations is not an option.

A sensing unit, as an integral part of sensor nodes, acquires information about a phenomenon of interest, such as proximity, acceleration,

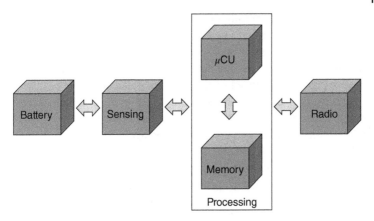

Figure 2.2 A typical wireless sensor node.

and smoke. On top of that, a sensing unit can also include various types of sensors, and thus it becomes able to collect data on multiple parameters simultaneously. Given that sensing units acquire physical data, the output becomes analog, which subsequently feeds the processing unit, where digital signals are processed. To this end, a sensing unit should include an analog-to-digital converter (ADC) along with the sensor to make its output suitable for the processing unit.

Even though sensor nodes are small devices with limited capabilities, they are able to process the acquired data locally with their in-built processing unit, which is composed of a micro-controller and memory. Moreover, as sensor nodes communicate with each other and with sink nodes, they also require a radio to enable wireless communications. Based on the network architecture, the communication between nodes can be either single-hop or multi-hop. While the former requires a direct link between the source and destination, the latter is formed in the case of the absence of a direct link between the two. In multi-hop communication, some sensor nodes may need to relay the data collected by other nodes, which are unable to connect to the sink node directly. Multi-hop communication becomes more frequent in mesh topologies as opposed to star topologies, in which each sensor node is likely to have a direct connection with the sink node Dargie and Poellabauer (2010).

Depending on the type of processing required, each of the three elements of WSNs (sensor node, sink node, and end-user) is capable of data processing with different computational power. Sensor nodes, for instance, have typically limited computational powers; hence it would not be convenient to carry out computationally intense tasks. As mentioned in Dargie and Poellabauer (2010), in order to reduce the amount of data to be transmitted to a sink node, sensor nodes may need to process the data when they serve as a relay node. For example, they may filter out redundant data, where redundancy may be caused by the proximity of the sensors, hence generating highly correlated measurements Kalaivaani and Krishnamoorthy (2018).

Data reduction methods, such as compression and prediction, play an important role in energy efficiency, because a higher amount of data results in increased energy consumption due to a higher computational complexity and a larger communication volume among the source and destination. Therefore, in Section 2.4, we will elaborate this aspect by focusing on how the amount of data to be transferred and/or processed can be reduced.

2.3 Industrial WSNs

As previously mentioned, today's competitive business environment forces industries to modernize themselves with new technologies and advancements, since the risk of lagging behind would be too costly and difficult to recover from. To this end, there is a dominant trend of digitizing and automating industrial processes. SNs are a focal aspect of all digitization and automation efforts, as they enable remote monitoring and maintenance, and automated decision making. As explained above, wired SNs are challenging to deploy in the industrial environment for various reasons. Firstly, the harsh conditions in the industrial environment, such as extreme temperatures, humidity levels, and vibration, are not suitable for normal cable installations and would require additional precautions. Second, the laying of cables is costly and time consuming, and may require lengthy work interruptions. Moreover, the maintenance of wired SNs requires further interruptions and expenditure, thus increasing the operational cost. These challenges are obviously intensified in the case of wide area sensor deployment. Consequently, industries are opting for WSNs to remain competitive and agile Paavola and Leiviska (2010).

WSNs fit very well with industrial applications due to their numerous advantages including low installation and maintenance costs, mobility support, enabling self-organization, etc. Queiroz *et al.* (2017). Industrial WSNs (IWSNs) have already gained a lot attention in both industry and academia. What makes IWSNs different from other types of WSNs is that IWSNs are specially designed for the needs of industrial environments and applications, which have distinctive characteristics. In the following sections, IWSNs will be discussed with respect to their characteristics and challenges. Then, wireless communication standards for IWSNs will be introduced before presenting some selected applications.

2.3.1 Requirements and Challenges

As outlined in Gungor und Hancke (2009), IWSNs have specific requirements that should be carefully addressed in order to avoid any type of undesired outcome. Firstly, large industrial plants may require a huge amount of sensor nodes, which subsequently increases the overall cost and occupied area. Therefore, the deployed sensor nodes should be **compact** and **low-cost** in order to make the implementation feasible. Secondly, since the life-span of sensor nodes is constrained by the capacity of their batteries, energy consumption becomes a crucial issue in IWSNs; hence, **energy efficiency** is an integral part of making IWSNs sustainable.

Thirdly, industrial environments can change quite quickly due to various reasons, including the deployment of new equipment and node malfunctioning/replacements. In addition, wireless channels vary over time in industrial plants due to the inevitable high level of mobility and line of sight Queiroz *et al.* (2017). Thus, **self-organization** is greatly required in IWSNs in order to be agile and adaptive to changing conditions. Fourthly, as decision-makers, e.g. the field manager, rely on the data acquired by sensor nodes, any kind of fault in the node may lead to catastrophic reactions. For example, malfunctioning in a sensor node, which measures the temperature of a particular expensive piece of machinery to keep its temperature under a certain value, will result in severe damage due to inaccurate readings. Hence, **reliability** in the measured and transferred data is crucial in IWSNs.

Lastly, IWSNs should also be **secure**, as malignant attacks threaten the reliability of communication among nodes due to the characteristics of wireless networks Christin *et al.* (2010). This threat, on the one hand, may damage the business of a company by causing wrong decisions on important and costly issues. On the other hand, as the most important aspect, it may also cause fatal accidents in the workplace, making the lives of employees vulnerable.

2.3.2 Protocols and Standards

There are seven layers in an open systems interconnection (OSI) model:

1) Physical layer
2) Data-link layer
3) Network layer
4) Transport layer
5) Session layer
6) Presentation layer
7) Application layer.

Among these, layers 1 to 4 are referred as the lower layers, while layers 5 to 7 are called the upper layers.

As discussed in Paavola and Leiviska (2010), the physical layer, as the lowest level of the OSI model, is responsible for selecting the frequency to be used and for implementing the modulation and data encryption. In addition to error detection/correction, the data-link layer, which consists of medium access control (MAC) and logistic link control (LLC), is also responsible for accessing to a transmission medium to prevent interference among the nodes Paavola and Leiviska (2010).

Routing data packets among nodes is the responsibility of the network layer. The transport layer participates in the process if data is transmitted to the end-user through the internet, while the application layer encapsulates the upper layers and responsible for making the end-user aware of the lower layer implementations Flammini *et al.* (2009).

There are various wireless communication standards for IWSNs:

- IEEE 802.15.4 defines how low-rate wireless personal area networks (LR-WPANs) operate by specifying the physical layer and MAC. This is quite convenient for IWSNs, as it targets short-range, low-cost, and low-power communications Queiroz *et al.* (2017).

It constitutes a base for the following technologies that are being used in IWSNs:

- Zigbee is typically designed for devices using batteries as an energy supplier due to its low-power properties. In addition, Zigbee is compact and low-cost, supports low data-rates, and has limited coverage; hence it is a good fit for IWSNs.
- {WirelessHART} uses the 2.4 GHz industrial, scientific, and medical (ISM) radio band, and targets industrial applications, as it is developed specifically for automation purposes.
- ISA 100.11a was developed by the International Society of Automation (ISA) in 2009 by focusing on wireless systems for industrial automation Queiroz *et al.* (2017).
- IPv6 over Low-Power Wireless Personal Area Networks (6LoW-PAN) was designed for IP based communication for small devices, such as sensor nodes in WSNs. It enables IPv6 packet transmission in networks that use IEEE 802.15.4 standards. As mentioned in Gungor und Hancke (2009), it is quite beneficial in IWSNs, since IP devices can communicate with each other directly or through the internet.

• Wireless Networks for Industrial Automation - Process Automation (WIA-PA) is a wireless communication standard and was specifically designed for process automation in industry. Although the physical layer is inherited from IEEE 802.115.4, the protocol also includes data-link and network layers Zhong *et al.* (2010).

• Ultra wideband (UWB) is a part of short-range wireless communication and offers a large bandwidth, enabling many WSN applications, such as video surveillance and imaging of environments Zhang *et al.* (2009).

• Wi-Fi is not primarily designed for industrial applications, but offers an improved quality of service (QoS) with higher data rates. Therefore, as mentioned in Li *et al.* (2017), it requires some adaptations to comply with industrial requirements.

2.3.3 IWSN Applications

Since WSNs have been used in industrial environments for a long time, there are numerous IWSNs applications, as detailed by Erdelj *et al.* (2013). As such, in this section we will discuss some selected state-of-the-art industrial applications.

In Erdelj *et al.* (2013), the authors propose a taxonomy and classification for IWSNs applications, dividing them into three main branches:

- Environmental sensing
- Condition monitoring
- Process automation.

Environmental sensing in industry refers to monitoring and detecting the issues in the habitats of industrial plants, such as pollution, hazards, and security Erdelj *et al.* (2013). Another example is flooding, which occurs frequently in many areas, and it can affect (or even destroy) industrial processes if the plants are located around flood-risk areas. For that reason, detecting floods plays a vital role, and the authors in Mousa *et al.* (2016) studied the detection of flash floods with their introduced sensor. In addition to hazards, protection of products is one of the main issues in an industrial process, and thus securing the plants from break-ins is essential. As such, the authors in Benzerbadj *et al.* (2018) investigated a cross-layer communication protocol for a fence surveillance application.

Industries should also monitor the condition of their equipment to make their business plans more accurately. This kind of monitoring enables them to be proactive, meaning that they can forecast possible failures to prepare themselves accordingly. For example, a structural health monitoring system for civil infrastructures was studied in Hackmann *et al.* (2014) with their proposed cyber physical co-design approach.

Process automation is another significant application in industry since it can enhance the overall efficiency and effectiveness by monitoring the processes and services Erdelj *et al.* (2013). In this regard, a predictive supply chain management for perishable goods was presented in Annese and Venuto (2015), where the authors predict the shelf-life of foods using measured environmental phenomena.

2.4 Life-span Extension for WSNs

Energy is always the main determinant for the life-span of WSNs, as they typically use external batteries, which have limited capacities. Moreover, even a single node can determine the life-span of a whole network if its absence affects the working routine of the system; i.e. it can be a sink node aggregating and routing the data from sensor nodes to an end-user, or a relay node that is located at a very strategic location and relays data coming from multiple neighbouring nodes. Alternatively, it can be a regular sensor node that is acquiring very important information, which the whole system relies on. Therefore, it is crucial to improve the energy balance of each individual node to extend its life-span, which in turn extends the life-span of the whole network. Let the energy balance of a single node be as follows:

$$E_b = E_{in} - E_{out}, \tag{2.1}$$

where E_{in} and E_{out} are the input and output energies, respectively.

It is obvious from (2.1) that the energy balance of networks can be increased by either increasing the energy input or decreasing the energy output, or by doing both. Given that WSNs work with limited energy sources (batteries), increasing the energy input refers to using alternative energy sources, such as solar and wind, along with the available batteries. Methods that enable using these alternative energy resources are called *energy harvesting*.

On the other hand, decreasing the energy output refers to *energy conservation*, which implies using the available sources as efficiently as possible by avoiding any kind of energy wasting.

2.4.1 Energy Harvesting

Energy harvesting techniques play a crucial role in extending the life-span of WSNs, as they basically use alternative energy sources to contribute to the energy balance of networks. However, as this process is performed with ambient energy sources, the harvested energy should be converted to electrical energy so that it can be used in sensor nodes. A general schematic for a typical energy harvesting system provided by Sudevalayam and Kulkarni (2011) is summarized in Figure 2.3.

The energy source, harvesting methods, and load have been reported as the key elements of an energy harvesting system in Sudevalayam and Kulkarni (2011), in which energy sources are cate-gorized into controllable and non-controllable. Controllable energy sources are the ones that can always provide energy on-request, whereas non-controllable sources, such as solar and wind, have their

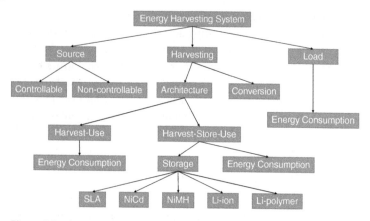

Figure 2.3 A schematic for a typical energy harvesting system.

on routine and may not be available when needed Sudevalayam and Kulkarni (2011); Tuna and Gungor (2016).

Moreover, there are two harvesting architectures available, as shown in Fig. 2.3 Sudevalayam and Kulkarni (2011): harvest-use and harvest-store-use. In the former, harvested energy is directly used in sensor nodes, while in the latter it is stored in a storage element, which is then used as an energy supplier by the sensor nodes.

The energy conversion technique to be employed depends on the type of energy source. In the case of solar energy, for instance, solar panels, which convert solar energy to electrical energy, should be employed. However, the solar panels generate direct current (DC), and thus a DC–DC converter might be needed in order to make the output of the panels suitable for a load, which is a sensor node in this case. Therefore, the DC–DC converter is supposed to be designed in an efficient manner, since it is more likely to result in a loss during conversion. Given that the conversion efficiency of the current solar panels are not sufficiently high, the loss in DC–DC converters also contributes negatively to the energy balance.

Various harvesting techniques are surveyed in both Sudevalayam and Kulkarni (2011) and Tuna and Gungor (2016) in a comprehensive manner. The overall list of the surveyed techniques is as follows:

- Solar energy harvesting
- Wind energy harvesting
- Radio frequency (RF) harvesting

- Piezoelectric energy harvesting
- Thermal energy harvesting.

Each of the harvesting techniques will be discussed in detail in the following sections.

2.4.1.1 Solar Energy Harvesting

Solar energy is non-controllable and one of the most common sources used for harvesting purposes. In solar energy harvesting, photovoltaic (PV) panels, which consist of multiple solar cells, take the responsibility of converting sunlight into electrical current. Referring to the single-diode circuit model of a solar cell, photo-generated current is modelled as follows De Soto *et al.* (2006):

$$I_{L} = \frac{S}{S_{ref}} \frac{M}{M_{ref}} \left[I_{L,ref} + \alpha_{I_{sc}} (T_c - T_{c,ref}) \right], \tag{2.2}$$

where S and S_{ref} are experienced and reference total absorbed irradiance, respectively; M and M_{ref} are the experienced and reference air mass modifiers, respectively; $I_{L,ref}$ is the reference light current; I_{sc} is the short-circuit current; T_c and $T_{c,ref}$ are the experienced and reference cell temperatures, respectively. Note that the reference parameters are calculated at standard rating conditions, where irradiance is 1000 W m^{-2} and the cell temperature is 25 deg C.

Equation (2.2) implies that the generated current by a solar cell is proportional to the experienced irradiation and temperature, which vary over time. As such, the harvest-store-use architecture presented in Sudevalayam and Kulkarni (2011) is convenient to use in solar energy harvesting systems.

There are various available solar energy harvesting nodes as provided in Sudevalayam and Kulkarni (2011), and we have identified the following more recently developed ones:

- EcoMicro Lee *et al.* (2018)
- OpenWise González *et al.* (2012)
- EscaCap Kim and Chou (2011).

2.4.1.2 Wind Energy Harvesting

Wind, as another environmental energy source, can also be used for energy harvesting purposes. The basic idea behind this technique is that wind causes circular motion in turbines, which is in turn converted into electrical energy.

The wind power can be calculated as Weimer *et al.* (2006):

$$P = \frac{1}{2}\rho A v^3, \tag{2.3}$$

where ρ is the air density, A is the rotor swept area, and v is the wind velocity. Therefore, it is quite clear from (2.3) that the velocity of wind is a significant parameter in the output power, as it contributes to the equation with its cube. In this regard, similar to the solar energy harvesting case in Section 2.4.1.1, energy storage elements are required to provide stable energy to the sensor nodes.

Although wind is a non-controllable energy source, its characteristics can be predicted for certain locations using historic wind data. In Wu *et al.* (2018), the authors tried to predict the energy harvesting for wind to avoid the effects of variations.

2.4.1.3 Radio Frequency Energy Harvesting

Since electromagnetic radiation is available almost everywhere in modern times, it is quite possible to exploit it to harvest energy. In RF energy harvesting techniques, an antenna is employed to receive ambient RF signals at certain frequencies. Then, the received RF signal is converted to a DC signal by utilizing a matching circuit and a rectifier, and thus the resulting power can be used to energize sensor nodes. In order to better assess the performance of rectifying antennas, which are basically the combination of the receiving antenna and the RF–DC converter, the concept of harvesting capacity is introduced in Vandelle *et al.* (2018).

One example of this is RF identification (RFID) systems. Unlike active ones, passive RFID tags do not include an in-built battery system; hence they need to harvest energy to power themselves. The electromagnetic field generated by an RFID reader is exploited by an RFID tag using magnetic coupling, which in turn creates mutual inductance Sudevalayam and Kulkarni (2011). Similar to this idea, RF energy harvesting can be an option for the electromagnetic radiations of TV broadcasting antennas, LTE/GSM base stations, etc.

2.4.1.4 Piezoelectric Energy Harvesting

The piezoelectric effect is a concept that is used to generate electric charge when mechanical stress is applied. Piezoelectric energy harvesting is quite applicable in industrial plants, where vibration is an ambient energy source due to machinery in operation. In order to energize the sensor nodes, the resulting vibration can be converted to electrical energy using piezoelectric converters, whose power density is around 500 μW cm^{-2} Tuna and Gungor (2016). The authors in Yang *et al.* (2018), for example, designed a piezoelectric windmill that works at low wind conditions, where it is inefficient to harvest energy with motors.

2.4.1.5 Thermal Energy Harvesting
The idea behind thermal energy harvesting is to exploit differences in temperature in order to generate electrical energy. There are two main types of thermal energy harvesting: thermoelectric and pyroelectric Basagni *et al.* (2013). As stated by Verma and Sharma (2019), in which the authors presented their algorithm for thermoelectric energy harvesting in WSNs, thermoelectric energy harvesting uses the principle of Seebeck effect:

$$\Delta V = \alpha \Delta T, \tag{2.4}$$

where α is material-specific Seebeck coefficient and T is temperature. As seen from (2.4), the generated voltage is directly proportional to the temperature difference and Seebeck coefficient. Therefore, the general idea is to have two dissimilar electrical conductors with different temperatures in order to create a potential difference.

In pyroelectric energy harvesting, on the other hand, changes in temperature over time, which alters the distribution of atoms in a crystal, is utilized instead of having a thermocouple. This kind of location changes of atoms subsequently generates electric voltage.

2.4.2 Energy Conservation

In order to decrease E_{out} in (2.1), the energy consumption of WSNs should be minimized to extend their life-span. Among the three main tasks (sensing, processing, and communicating) performed in a typical sensor node in WSNs, communication is the most energy consuming one Akyildiz *et al.* (2007). However, since sensor nodes are energy-constraint devices, it is crucial to reduce the energy consumption in each phase. Energy conservation methods in SNs are comprehensively surveyed in Anastasi *et al.* (2009), of which we will adopt the high-level taxonomy.

2.4.2.1 Duty Cycling

Duty cycling, in a broader perspective, embodies a collection of methods that target minimizing the energy consumption of SNs by reducing the amount of time that the nodes are in active mode, which is the case when the nodes are in the process of data transmission. Duty cycling methods are divided into two subcategories in Anastasi *et al.* (2009): *topology control* and *power management*.

While sensor deployments can be carried out in bulk and in a random manner, especially when an environment of interest is wide, it is quite possible to include redundant sensors, playing an insignificant role in maintaining the connectivity, in the network. Topology control refers to the techniques aiming to put redundant nodes in sleep mode in order to save energy. However, identifying the redundant nodes and finding an optimum network topology, in which each active node is an integral part of ensuring the connectivity throughout the network, is a challenging task Anastasi *et al.* (2009).

In Elhabyan *et al.* (2018), the authors presented a coverage control mechanism that is responsible for finding the optimal set of active nodes without compromising on coverage. They employed a genetic algorithm, and the proposed methodology is easily appended to centralized clustering protocols.

In power management, the focus becomes the active nodes rather than the redundant ones. The main goal is to reduce the energy consumption of active nodes by putting their radio in sleep mode when they do not have data to transmit, and they remain in sleep mode until data transmission is needed Anastasi *et al.* (2009). By doing so, the duty cycle of active nodes decreased, which in turn conserves some energy. In Kubota *et al.* (2017), for example, the authors introduced a sleep/wake-up protocol for nodes in order to reduce the energy consumption. It is also reported in their study that the proposed protocol is capable of decreasing the activity of radio by around 90%. The authors in Nadas *et al.* (2016) studied alarm based WSNs by investigating the relationship between the energy consumption arising from more frequent time synchronization and the energy conservation due to decreasing the idle listening window. The optimum number of synchronization was derived analytically, and they achieved a significant energy conservation with their proposed methodology.

2.4.2.2 Data Driven Approaches

Duty cycling brings a broader idea to minimize the energy consumption: put the nodes in sleep mode if they are redundant or do not have data to send. Data driven approaches, in contrast, focus on the energy consumption during data processing and/or transmitting. In other words, there is still room for energy conservation while a node is active and processing/transmitting data. Moreover, the volume of energy consumption is directly proportional to the amount of data being processed/transmitted Ozturk *et al.* (2018a).

In this regard, the subsets of the data driven energy reduction concept are reported in Anastasi *et al.* (2009) and are divided into:

- In network processing
- Data compression
- Data prediction
- Adaptive sampling
- Hierarchical sampling
- Model-driven active sampling.

Even though each of these methods have different approaches, their common ground is to minimize the amount of data to be processed/transmitted.

In the data prediction method, for example, the focus is the volume of data transmission between source and sink nodes. A model of sensed data is stored at the sink node so that it can predict the data without requesting the actual measurements form the source node. This, in turn, decreases the volume of the communication between the two. In this process, the source node always validates the model available at the sink node by comparing it with the actual measurements, and the model is updated if it is no longer valid Anastasi *et al.* (2009). This kind of prediction-based model is proposed in Dinh and Kim (2019) by considering both the prediction accuracy and energy efficiency. Based on the accuracy requirements of applications, the developed model is able to adjust the number of active sensors and the timing of data transmission.

2.4.2.3 Mobility Based Approaches

Since the methods given in Sections 2.4.2.1 and 2.4.2.2 do not consider the problems arising from locations of the deployed sensor nodes, there is still room for improving the energy efficiency in WSNs. In the mobility-based energy conservation approaches, a mobile data collector, which travels around the network and collects data from the sensor nodes, is employed. These data collectors (sink or relay) can be the part of either the network (controllable), or environment (non-controllable) Anastasi *et al.* (2009).

The benefits of having a mobile data collector are two-fold:

1) It decreases the burden of the nodes that have a close proximity to sink nodes, since they mostly need to serve as a relay node, which subsequently drains their battery. This works with the fact that the life-span of a single node can determine the life-span of a whole network if it is responsible for a significant task or is important in maintaining the connectivity. The mobile data collectors can alter the data flow and relive the load on some particular nodes, which in turn makes them consume less energy in the data relaying duty.
2) It decreases the probability of link errors caused by multi-hop communication by minimizing the number of required hops.

The authors of Cheng and Yu (2018) investigated the case of mobile relay by considering not only the energy consumption but also the delay originating in the data gathering process. They also considered the data loss problem by restricting the number of assisting sensor nodes. Non-controllable mobile data collectors, such as buses, are very likely to have predictable profiles due to experienced traffic conditions, which often include daily and weekly motifs. Thus, it could be a good solution to predict the locations of the mobile data collectors in order to better arrange the data gathering process. In this regard, a mobility prediction method as in Ozturk *et al.* (2018b) can be adopted, even though the authors considered cellular mobile networks.

2.4.2.4 *Q* Learning Assisted Energy Efficient Smart Connectivity

It has been a truism that integration of machine learning algorithms to wireless networks has great impacts on improving their performance, as they enable wireless networks to be more dynamic, agile, and versatile Ozturk *et al.* (2019). Reinforcement learning is one branch of machine learning, and has been applied to various domains for optimization purposes. Problems requiring a smart decision making are especially good fits for reinforcement learning. In this regard, in Ozturk *et al.* (2018a), we implemented a *Q* learning driven solution for IoT networks to select the wireless connection type and data processing location by considering both the application requirements and the end-to-end energy consumption.

Q learning is an algorithms in reinforcement learning and offers great opportunities in optimizing wireless networks. Moreover, it has already been proven that Q learning is capable of working properly in dynamic environments Ozturk *et al.* (2018a). The main components of a typical Q learning framework are *agent*, *action*, *states*, *action-value function*, and *reward/penalty*. In short, the agent takes actions and then evaluates the reward/penalty and the corresponding state in order to update the action-value function with the following backup:

$$Q(s_t, a_t) \leftarrow Q(s_t, a_t) + \lambda \big[R_{t+1} + \alpha \cdot \max_a (Q(s_{t+1}, a)) - Q(s_t, a_t) \big],$$

(2.5)

where s_t and s_{t+1} are the states at time t and $t + 1$, respectively; a_t and a are the action taken and the set of all possible actions, respectively; R_{t+1} is the expected reward; λ and α are the learning rate and discount factor, respectively. Note that R in (2.5) can be replaced with penalty, P, in the case of penalty based implementation, where the max function is also replaced with min function.

Q learning is also a model-free algorithm Sutton and Barto (2018), meaning that it does not require a model for an environment of interest in order to learn its optimal behaviour; rather it interacts with the environment and determines its policies accordingly. It is also an off-policy method, which implies that it follows different policies in determining the action to be taken and in updating the action-value function. Although an ϵ-greedy can be the base policy, it follows a π policy, where $\epsilon > 0$, in determining the action, while a μ policy, where $\epsilon = 0$, is followed in updating the action-value function.

In this context, in Ozturk *et al.* (2018a), we considered a smart-port scenario, where IoT devices have multiple options for connection and data processing location. To this end, we assumed that Wi-Fi and narrowband IoT (NB-IoT) are available for wireless connection, and local, fog, and cloud processing options are available for data processing. Furthermore, since each application that IoT devices run might have different demands, we identified three application-specific requirements: *response time*, *security*, and *data rate*. The prime objective of Ozturk *et al.* (2018a) is to select the optimum pair of wireless connections (Wi-Fi and NB-IoT) and data processors (local, fog, and cloud) for a given application in view of each connection and processing option having distinctive characteristics. If, for example, the connection is required to be as secure as possible, then the choice should be NB-IoT, as it is more secure than Wi-Fi owing to the eSIM card.

We proposed a two-stage process, in which the first stage decides the pair of wireless connection and processing location by considering the requirements of the applications as well as the energy consumption, while the second stage is executed if only the decision on the first stage includes fog or cloud processing. Thus, in the second stage, the percentage of the data to be offloaded to the fog or cloud is determined with the help of Q learning by taking into account the capacity of the fog/cloud and the available budget for data processing. Note that there is no fog processing option when NB-IoT is selected as a wireless connection, since it directly connects IoT devices to the evolved node Bs (eNBs) without requiring a gateways as in Wi-Fi.

In terms of energy consumption for data processing: $E_{p,d} > E_{p,f} > E_{p,c}$, where $E_{p,d}$, $E_{p,f}$, and $E_{p,c}$ are the energy consumption for local, fog, and cloud processing, respectively. However, this is the opposite for data processing charges: $C_{p,c} > C_{p,f} > C_{p,c}$, where $C_{p,d}$, $C_{p,f}$, and $C_{p,c}$ are the monetary costs for local, fog, and cloud processing, respectively.

The proposed model in Ozturk *et al.* (2018a) aims to satisfy the IoT devices with their requirements while reducing their energy consumption. In other words, the solution tries to select the optimum connection and processing pair that satisfies the application-specific requirements while resulting in less energy consumption.

We compared the proposed model with six different benchmark scenarios with different wireless connection types and data processing locations. First, 10 IoT devices are assumed to exist in the coverage area of a gateway. Then, these devices are equally grouped into two according to their wireless connection types: the ones with NB-IoT constitute group I, while group II includes the devices using a Wi-Fi connection. The data processing location for each device is shown in Table 2.1, in which all the possibilities are included.

Figure 2.4 demonstrates the performances of the benchmark methods compared to the proposed method. The methods are compared according to following key performance indicators (KPIs): energy consumption, monetary cost, satisfaction of IoT devices, number of out of budget IoT devices, and a holistic penalty, which is the

Table 2.1 List of the fixed scenarios with data processing locations and connection types.

Scenario	Group I	Group II
A	Device	Device
B	Cloud	Device
C	Device	Fog
D	Cloud	Fog
E	Device	Cloud
F	Cloud	Cloud

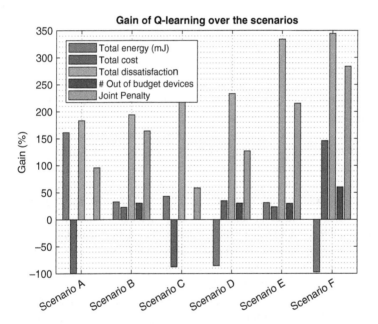

Figure 2.4 Proposed Q learning assisted two-stage model.

cumulative combination of all four KPIs. The obtained results reveal that the proposed model outperforms the benchmark scenarios significantly in terms of the holistic penalty with values ranging from 95.9% and 283.54%. This results prove the applicability of the Q learning algorithm to multi-objective problems.

In terms of monetary cost, as can be seen, it is quite hard to outperform scenarios A and C, since they are the ones with minimum cost due to their device and fog computations. Similarly, scenarios D and F include cloud and fog processing, and as such their energy consumption is the minimum, and thus it is not possible to outperform them in terms of energy consumption. However, the proposed method manages to outperform scenarios A and C in all KPIs other than the monetary cost, and it outperforms scenarios D and F in all KPIs other than the energy consumption. Furthermore, the developed model gives better results than scenarios B and E in all KPIs. In summary, the proposed methodology is capable of satisfying the requirements of IoT devices (it managed to outperform all the scenarios in user satisfaction), while keeping the energy consumption as low as possible.

2.5 Conclusion

WSNs have become an integral part of industry in view of their numerous superiorities over wired networks; industrial plants are mostly harsh environments, making the installation and maintenance of wired networks very inefficient in terms of cost and time. WSNs render the industrial processes more dynamic and agile owing to their real-time monitoring capabilities, hence reducing the operational expenditure with minimal installation cost. However, the life-span of WSNs should be as long as possible in order for the industry to capitalize on the offered gain and maximize their cost effectiveness. Energy consumption is the main determinant factor for the life-span of WSNs, since the sensor nodes are typically energy constraint devices due to their limited batteries.

In this chapter, first we introduced WSNs in general, and focused on IWSNs by presenting their distinctive characteristics and some selected applications. Then, energy harvesting and energy conservation are identified as two methods to extend the life-span of WSNs. Various techniques for each method were introduced and the related state-of-the-art was provided. Lastly, we demonstrated our Q learning assisted energy efficient model, and provided the simulation results revealing that it is capable of improving the application specific gains of the WSN while reducing the energy consumption.

Bibliography

Akyildiz, I.F., Melodia, T., and Chowdhury, K.R. (2007) A survey on wireless multimedia sensor networks. *Computer networks*, 51 (4), 921–960.

Aliu, O.G., Imran, A., Imran, M.A., and Evans, B. (2013) A survey of self organisation in future cellular networks. *IEEE Communications Surveys Tutorials*, 15 (1), 336–361, doi:10.1109/SURV.2012. 021312.00116.

Anastasi, G., Conti, M., Di Francesco, M., and Passarella, A. (2009) Energy conservation in wireless sensor networks: A survey. *Ad hoc networks*, 7 (3), 537–568.

Annese, V.F. and Venuto, D.D. (2015) On-line shelf-life prediction in perishable goods chain through the integration of WSN technology with a 1st order kinetic model, in *Proc. IEEE 15th Int. Conf. Environment and Electrical Engineering (EEEIC)*, S. 605–610, doi:10.1109/EEEIC.2015.7165232.

Basagni, S., Naderi, M.Y., Petrioli, C., and Spenza, D. (2013) Wireless sensor networks with energy harvesting. *Mobile Ad Hoc Networking: Cutting Edge Directions*, S. 701–736.

Benzerbadj, A., Kechar, B., Bounceur, A., and Hammoudeh, M. (2018) Surveillance of sensitive fenced areas using duty-cycled wireless sensor networks with asymmetrical links. *Journal of Network and Computer Applications*, 112, 41–52.

Cheng, C. and Yu, C. (2018) Mobile data gathering with bounded relay in wireless sensor networks. *IEEE Internet of Things Journal*, 5 (5), 3891–3907, doi:10.1109/JIOT.2018.2844680.

Christin, D., Mogre, P.S., and Hollick, M. (2010) Survey on wireless sensor network technologies for industrial automation: The security and quality of service perspectives. *Future Internet*, 2 (2), 96–125.

Dargie, W. and Poellabauer, C. (2010) *Fundamentals of wireless sensor networks: theory and practice*, John Wiley & Sons.

De Soto, W., Klein, S., and Beckman, W. (2006) Improvement and validation of a model for photovoltaic array performance. *Solar energy*, 80 (1), 78–88.

Dinh, N. and Kim, Y. (2019) An energy efficient integration model for sensor cloud systems. *IEEE Access*, 7, 3018–3030, doi:10.1109/ ACCESS.2018.2886806.

Du, Y.C., Lee, Y.Y., Lu, Y.Y., Lin, C.H., Wu, M.J., Chen, C.L., and Chen, T. (2011) Development of a telecare system based on zigbee mesh network for monitoring blood pressure of patients with hemodialysis in health care centers. *Journal of Medical Systems*, 35 (5), 877, doi:10.1007/s10916-010-9513-0. URL http://dx.doi.org/10.1007/s10916-010-9513-0.

Elhabyan, R., Shi, W., and St-Hilaire, M. (2018) Evolutionary-based coverage control mechanism for clustered wireless sensor networks, in *International Conference on Wired/Wireless Internet Communication*, Springer, S. 67–80.

Erdelj, M., Mitton, N., Natalizio, E. *et al.* (2013) Applications of industrial wireless sensor networks. *Industrial Wireless Sensor Networks: Applications, Protocols, and Standards*, S. 1–22.

Fernández, M., Thiel, D.V., Arrinda, A., and Espinosa, H.G. (2018) An inward directed antenna for gastro-intestinal radio pill tracking at 2.45 GHz. *Microwave and Optical Technology Letters*, 60 (7), 1644–1649.

Flammini, A., Ferrari, P., Marioli, D., Sisinni, E., and Taroni, A. (2009) Wired and wireless sensor networks for industrial applications. *Microelectronics journal*, 40 (9), 1322–1336.

González, A., Aquino, R., Mata, W., Ochoa, A., Saldaña, P., and Edwards, A. (2012) Open-wise: A solar powered wireless sensor network platform. *Sensors*, 12 (6), 8204–8217.

Gungor, V.C. and Hancke, G.P. (2009) Industrial wireless sensor networks: Challenges, design principles, and technical approaches. *IEEE Transactions on Industrial Electronics*, 56 (10), 4258–4265, doi:10.1109/TIE.2009.2015754.

Hackmann, G., Guo, W., Yan, G., Sun, Z., Lu, C., and Dyke, S. (2014) Cyber-physical codesign of distributed structural health monitoring with wireless sensor networks. *IEEE Transactions on Parallel and Distributed Systems*, 25 (1), 63–72, doi:10.1109/TPDS.2013.30.

Hodge, V.J., O'Keefe, S., Weeks, M., and Moulds, A. (2015) Wireless sensor networks for condition monitoring in the railway industry: A survey. *IEEE Transactions on Intelligent Transportation Systems*, 16 (3), 1088–1106, doi:10.1109/TITS.2014.2366512.

Jawad, H., Nordin, R., Gharghan, S., Jawad, A., and Ismail, M. (2017) Energy-efficient wireless sensor networks for precision agriculture: A review. *Sensors*, 17 (8), 1781.

Kalaivaani, P. and Krishnamoorthy, R. (2018) Performance analysis of various hierarchical routing protocols using spatial correlation. *Measurement*.

Kim, S. and Chou, P.H. (2011) Energy harvesting by sweeping voltage-escalated charging of a reconfigurable supercapacitor array, in *Proc. IEEE/ACM Int. Symp. Low Power Electronics and Design*, S. 235–240, doi:10.1109/ISLPED.2011.5993642.

Kolavennu, S. and Gonia, P. (2016) Wireless gas sensors for industrial life safety, in *Industrial Wireless Sensor Networks*, Elsevier, S. 155–166.

Kubota, H., Teramae, J., and Wakamiya, N. (2017) An efficient sleep/wake-up protocol for localization in impulse wireless sensor networks, in *Proc. 23rd Asia-Pacific Conf. Communications (APCC)*, S. 1–6, doi:10.23919/APCC.2017.8304059.

Lee, C.T., Liang, Y.H., Chou, P.H., Gorji, A.H., Safavi, S.M., Shih, W.C., and Chen, W.T. (2018) Ecomicro: A miniature self-powered inertial sensor node based on bluetooth low energy, in *Proceedings of the International Symposium on Low Power Electronics and Design*, ACM, S. 30.

Li, X., Li, D., Wan, J., Vasilakos, A.V., Lai, C.F., and Wang, S. (2017) A review of industrial wireless networks in the context of industry 4.0. *Wireless networks*, 23 (1), 23–41.

Low, K.S., Win, W.N.N., and Er, M.J. (2005) Wireless sensor networks for industrial environments, in *null*, IEEE, S. 271–276.

Mousa, M., Zhang, X., and Claudel, C. (2016) Flash flood detection in urban cities using ultrasonic and infrared sensors. *IEEE Sensors Journal*, 16 (19), 7204–7216, doi:10.1109/JSEN.2016.2592359.

Nadas, J.P.B., Souza, R.D., Pellenz, M.E., Brante, G., and Braga, S.M. (2016) Energy efficient beacon based synchronization for alarm driven wireless sensor networks. *IEEE Signal Processing Letters*, 23 (3), 336–340, doi:10.1109/LSP.2016.2515580.

Ozturk, M., Gogate, M., Onireti, O., Adeel, A., Hussain, A., and Imran, M.A. (2019) A novel deep learning driven low-cost mobility prediction approach for 5G cellular networks: The case of the control/data separation architecture (CDSA). *Neurocomputing*, doi:https://doi.org/10.1016/j.neucom.2019.01.031.

Ozturk, M., Jaber, M., and Imran, M.A. (2018a) Energy-aware smart connectivity for IoT networks: Enabling smart ports. *Wireless Communications and Mobile Computing*, 2018.

Ozturk, M., Klaine, P.V., and Imran, M.A. (2018b) Introducing a novel minimum accuracy concept for predictive mobility management schemes, in *Proc. IEEE Int. Conf. Communications Workshops (ICC Workshops)*, S. 1–6, doi:10.1109/ICCW.2018.8403507.

Paavola, M. and Leiviska, K. (2010) Wireless sensor networks in industrial automation, in *Factory Automation*, InTech.

Queiroz, D.V., Alencar, M.S., Gomes, R.D., Fonseca, I.E., and Benavente-Peces, C. (2017) Survey and systematic mapping of industrial wireless sensor networks. *Journal of Network and Computer Applications*, 97, 96–125.

Sudevalayam, S. and Kulkarni, P. (2011) Energy harvesting sensor nodes: Survey and implications. *IEEE Communications Surveys Tutorials*, 13 (3), 443–461, doi:10.1109/SURV.2011.060710.00094.

Sutton, R.S. and Barto, A.G. (2018) *Reinforcement learning: An introduction*, MIT press.

Tuna, G. and Gungor, V. (2016) Energy harvesting and battery technologies for powering wireless sensor networks, in *Industrial Wireless Sensor Networks*, Elsevier, S. 25–38.

Vandelle, E., Bui, D.H.N., Vuong, T.P., Ardila, G., Wu, K., and Hemour, S. (2018) Harvesting ambient RF energy efficiently with optimal angular coverage. *IEEE Transactions on Antennas and Propagation*, S. 1, doi:10.1109/TAP.2018.2888957.

Verma, G. and Sharma, V. (2019) A novel thermoelectric energy harvester for wireless sensor network application. *IEEE Transactions on Industrial Electronics*, 66 (5), 3530–3538, doi:10.1109/TIE.2018.2863190.

Weimer, M.A., Paing, T.S., and Zane, R.A. (2006) Remote area wind energy harvesting for low-power autonomous sensors, in *Power Electronics Specialists Conference, 2006. PESC'06. 37th IEEE*, IEEE, S. 1–5.

Wu, Y., Li, B., and Zhang, F. (2018) Predictive power management for wind powered wireless sensor node. *Future Internet*, 10 (9), 85.

Yang, C.H., Song, Y., Jhun, J., Hwang, W.S., Do Hong, S., Woo, S.B., Sung, T.H., Jeong, S.W., and Yoo, H.H. (2018) A high efficient piezoelectric windmill using magnetic force for low wind speed in wireless sensor networks. *Journal of the Korean Physical Society*, 73 (12), 1889–1894.

Zhang, J., Orlik, P.V., Sahinoglu, Z., Molisch, A.F., and Kinney, P. (2009) UWB systems for wireless sensor networks. *Proceedings of the IEEE*, 97 (2), 313–331.

Zhong, T., Mengjin, C., Peng, Z., and Hong, W. (2010) Real-time communication in WIA-PA industrial wireless networks, in *Computer science and information technology (ICCSIT), 2010 3rd IEEE international conference on*, Bd. 2, IEEE, Bd. 2, S. 600–605.

3

Multiple Access and Resource Sharing for Low Latency Critical Industrial Networks

Mohsin Raza[1], Anas Amjad[2], and Sajjad Hussain[3]

[1] *Faculty of Science and Technology, Middlesex University, UK*
[2] *Staffordshire University, UK*
[3] *James Watt School of Engineering, University of Glasgow, UK*

3.1   Introduction

The communications between sensors and control units play an important role in the smooth running of industry. Due to the utmost importance of communications within different blocks of industry, a lot of research can be witnessed in this domain. With the rapid advancements in industry, communication systems have also evolved over the years. Presently, industrial wireless sensor networks (IWSNs) offer several standards to deal with diverse applications. The standards and research developments have resulted in improved performance in different industrial systems (Raza et al., 2018a).

IWSNs provide infrastructure to offer suitable solutions to deal with the ever-changing requirements of industrial systems. The ongoing developments in IWSNs and research findings have opened up new venues for improvements within industry. The coexistence of diverse industrial processes and requirements for low latency communications offer many challenges and suitable solutions are essential.

The critical industrial applications require low latency communications for effective management of automation, process control and feedback systems (Raza et al., 2017). The diverse nature of applications within the industries and the varying time and reliability

Wireless Automation as an Enabler for the Next Industrial Revolution, First Edition.
Edited by Muhammad A. Imran, Sajjad Hussain and Qammer H. Abbasi.
© 2020 John Wiley & Sons Ltd. Published 2020 by John Wiley & Sons Ltd.

requirements of individual applications demand nonlinear and precedence based information scheduling and communications for better optimization of the critical processes. In IWSNs, multiple access techniques provide means to effectively manage and share the resources depending on the needs and criticality of an application. In industrial systems, since some applications might be more critical than others, effective and on-demand resource sharing allows overcoming the latency and reliability requirements.

In this chapter, the traffic within industrial systems is discussed in detail. It covers a detailed review of various medium access control (MAC) protocols. In addition, a MAC emergency communications scheme is presented to address challenges within industrial communications.

The recent developments in IWSNs play an important role in industrial evolution and performance improvement in a wide range of industrial applications. Due to the broader incorporation of automated processes within industry, the importance of sensory data and its timely communication has become very important. The communications infrastructure in industries has been thoroughly revised to support high speed communication between the sensors and decision blocks/ control units. The communications architecture has also been transformed to support low latency communications within different blocks of industrial systems (Kan et al., 2013). Further to this, suitable changes have been introduced to incorporate emergency and highly critical communication on priority bases within the industries. The close vicinity of diverse applications and processes within the industries also introduce information scheduling and transmission problems within IWSNs. In addition, strict time deadlines also pose certain challenges.

Communication within industrial systems can be divided into different classes based on the requirements of different applications. To facilitate diverse communication requirements in IWSNs, different channel access schemes and architectural changes have been promoted. The data and traffic generated within the industry can be divided into several classes based on the industrial system under consideration.

The main concerns of IWSNs in terms of data communications mainly arise from traffic originating from the critical processes within the industries. The critical data traffic generated by the industry mainly deals with emergency, regulatory, and supervisory control systems. Due to the high priority of such processes and relatively shorter time deadlines, priority based or dedicated access is provided to traffic from the above mentioned industrial systems.

Within IWSNs, the traffic is divided into six classes, including emergency/safety traffic, regulatory control traffic, supervisory control traffic, open-loop control traffic, alerting traffic and monitoring traffic (Raza et al., 2018a). Each of these traffic types are briefly explained as follows.

1) Safety or emergency systems are the backbone of industrial systems and ensure safe running of the processes. Such traffic is usually generated within the industries if some critical processes have been endangered or potentially harmful incidents have been recorded that can threaten the equipment as well as human life. The traffic generated by emergency systems has highest priority and need to be reported instantly. Usually communications of emergency traffic in IWSNs is facilitated by a dedicated communication/pilot channel with prioritized slot access (Gungor and Hancke, 2009).

2) Regulatory control traffic is essential for smooth running of the processes. Delay or loss of such information may result in compromising the quality of the end product or ruining the preparation of the batch. In addition, delay in regulatory control traffic may also affect the efficiency of the processes and may result in inducing higher production costs. Regulatory control traffic has high reliability requirements and is communicated using guaranteed slotted access or prioritized slotted contention schemes (Raza et al., 2018a).

3) Supervisory control traffic is relatively less critical compared to emergency/safety and regulatory control traffic. The communication requirements are asynchronous in nature and communication is triggered if the thresholds are violated. Thus, supervisory control traffic can be divided in critical and non-critical. Usually slotted access is used to support communications of supervisory traffic.

Open loop traffic is usually used for the asynchronous communication of sensory data for open loop control systems. The communications of open loop traffic are usually informational and have limited impact on the processes. In some cases though, the communications of open loop traffic is forwarded to control centre where it is used for manually controlling the actuators within the processes. It has relatively lesser reliability requirements and occasional packet misses can be tolerated easily (Raza et al., 2018a).

Alerting traffic offers means to raise the alarm to the circumstances which have minimal impact on the current running processes. However, if the traffic generated by alerting systems is delayed beyond reasonable periods, this may result in the development of certain operational hazards. The communications have lower reliability requirements and frequent communication failure can be tolerated without any consequence. Usually slotted contention based channel access is used to relay alerting traffic to the control centre.

Monitoring traffic is mainly targeted to collect survey data from the industrial processes that have no direct consequence on the running processes. However, in many cases the data collected from monitoring systems is used for later developments of processes in the future. This traffic is mainly logged into the storage and collected over a longer duration of time to suggest suitable changes in the industrial processes that can result in overall performance improvement in the future. The monitoring traffic has the lowest priority and low reliability requirements. Therefore, best effort services are used for communications of monitoring data with the focus on improving the lifetime and energy efficiency of sensor networks (Raza et al., 2018a). The priority level and time constraints for these traffic types within the industry are presented in Figure 3.1.

The coexistence of diverse traffic types with strict performance limitations (reliability, latency) introduce several challenges. In order to offer a viable solution to ensure desired reliability, without violating strict time deadlines, various solutions have been proposed. In addition, to deal with the coexistence of different traffic types different scheduling algorithms have also been proposed. The upcoming sections discuss in detail the proposed solutions and offer a viable priority based communications infrastructure to address the reliability and latency challenges in IWSNs. In addition, static scheduling is also presented to facilitate the coexistence of diverse processes and traffic types without permanent resource allocations.

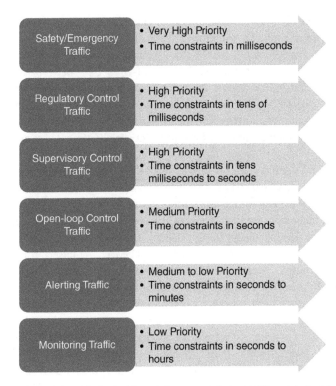

Figure 3.1 Priority levels and time constraints of different industrial traffic types.

3.2 Research Developments

Most of the developments on IWSNs use the physical layer attributes defined by IEEE 802.15.4 (IEEE Standard for Information Technology, 2006). In addition, several MAC protocols are also introduced to deal with the stringent demands of industrial systems and critical traffic. Due to the vast scope of MAC responsibilities, developments in MAC protocols have been very influential in ensuring effective communication strategies within the industries. Over the years, these protocols have adopted different multiple access schemes to suit the unique requirements of different industrial applications.

The recent developments in MAC layers proposed by the research community targeted improvements in reliability, real time data communications, energy efficiency and deterministic network formation.

Since the MAC layer handles access to the wireless medium and manages the use of radio, therefore suitable changes in MAC can lead to notable improvements within the performance of the protocols whether they are targeting reliability, real time data delivery or energy efficiency. In addition, the channel congestion can be better managed with systematic use of radio among individual nodes. Since the management of radio through the MAC layer also allows adaptive power management, thus ensuring longer network lifetime (Pei et al., 2013).

Due to the significance of the MAC layer, the MAC based protocols have notable significance and over the years a large number of MAC protocols have been proposed. These protocols can be classified into a number of categories. Some of these categories can be based on channel access, where MAC protocols are divided into carrier sense multiple access/collision avoidance (CSMA/CA) based, time division multiple access (TDMA) based, and multi-channel and hybrid schemes. Another division labels MAC protocols on the basis of the communication periodicity where MAC protocols are divided in asynchronous, synchronous and hybrid. Some other categories include, asynchronous, synchronous, slotted, hybrid priority based and multi-channel.

Since in this chapter, low latency industrial networks are considered; therefore the MAC protocols targeting energy efficiency and best effort communications are beyond the scope of discussion. Further details related to these MAC protocols can be found in (Demirkol et al., 2006; Naik and Sivalingam, 2004). In addition to this, a detailed survey of energy efficient MAC protocols is presented in (Roy and Sarma 2010–Kumar et al. 2018).

In the last decade a large number of MAC protocols were proposed. In Langendoen (2008) the authors divide the MAC protocols into periodic, slotted, random and hybrid. In Pei et al. (2013) the authors classified the MAC protocols into asynchronous, synchronous, slotted and multi-channel schemes.

Each of the sub-classes of the MAC protocols offer unique benefits along with certain limitations. Asynchronous MAC protocols communicate rarely and thus are more suitable to conserve large amounts of energy for longer network lifetime. Since such protocols are beyond the scope of discussion at this point, interested readers can refer to Doudou et al. (2013) where extended details of asynchronous protocols are presented. Since the asynchronous protocols are more suitable for monitoring applications, thus pose certain communication challenges in industrial applications. These challenges presented in asynchronous MAC protocols were addressed in synchronous protocols; however, in synchronous protocols, congestion and collision avoidance still remained a challenge. To overcome congestion and collision issues in asynchronous, synchronous/slotted schemes TDMA based channel access was used to provide guaranteed channel access. While TDMA resolves the congestion and collision issues, the predefined slotted access in TDMA introduced added delay. To reduce the delay, multi-channel schemes were also presented but due to the existence of TDMA the delay was still unacceptable for critical industrial applications, especially when multiple traffic types coexist.

Apart from these, there exist several protocols that target MAC optimization for both CSMA/CA and TDMA based channel access to tackle both periodic and emergency communications. In this chapter, MAC protocols are classified based on the channel access and, therefore, are divided in CSMA/CA based, TDMA based, multi-channel based and priority based. Each of these are listed as follows.

3.2.1 CSMA/CA Based MAC Schemes

A number of MAC protocols can be classified as CSMA/CA based medium access protocols. However these protocols do not provide deterministic behaviour. Due to packet collision and channel congestion, there exists a possibility that the packet may be lost or delayed. Therefore, these schemes are not suitable for critical industrial applications. However, due to the demand driven channel access in CSMA/CA based protocols, such schemes are more suitable for less critical industrial applications where the benefits of long-term batteries can be received. A critical analysis of the CSMA/CA based MAC protocols is presented in Anastasi et al. (2011). Some other contention based schemes that use a variety of techniques to improve packet transmission, information delay, and power utilization are presented in (Pangun et al. 2013–Marchenko et al. 2014).

3.2.2 TDMA Based MAC Schemes

TDMA based MAC protocols are more effective in providing reliable communications with bounded delays. The guaranteed channel access in TDMA based MAC protocols allow a wider acceptability of TDMA based channel access schemes in comparison to CSMA/CA based schemes. Therefore, a much wider adoptability of TDMA based channel access schemes can be witnessed in regulatory and open loop control systems. However, the time synchronization appears to be a major challenge in these protocols, especially in mesh and tree networks. The problem is less prominent in tree topology. In any case, TDMA based schemes require time synchronization for effective communication (Dobslaw et al., 2014; Ergen and Varaiya, 2010). There are few proposals that address the synchronization issues. In (Ergen and Varaiya, 2010), the authors proposed two heuristic algorithms to solve scheduling problems. The paper has also evaluated upper bounds on these schedules as a function of total packets generated in the network and ensured packet delivery. In (Shen et al., 2013; Yan et al., 2014), the results of (Ergen and Varaiya, 2010) were further improved where the new protocols claimed to improve the performance of (Ergen and Varaiya, 2010). The authors of (Shen et al., 2013) evaluated the results while considering a harsh dynamic environment; however the packet reliability suffered. The authors of (Yan et al., 2014) formed a hypergraph to improve scheduling flexibility and reliability in harsh environments. In addition, this paper also presented two schemes, i.e. dedicated and shared scheduling for performance improvement. ShedEx is another scheme that uses TDMA based channel access (Dobslaw et al., 2014). This paper improves reliability by replicating most rewarding slots and offer a specified reliability through scheduling algorithm.

3.2.3 Multichannel MAC Schemes

The use of multi-channel schemes allows improved channel utilization where the performance and capacity of the TDMA schemes is improved. A relatively higher number of TDMA based multi-channel MAC schemes have been introduced in the past few years. The authors of (Zhao et al., 2014) proposed a multi-channel TDMA based source aware scheduling scheme for static networks. The proposed schemes use multiple channels, but fail to offer reliability. The authors of (Dobslaw et al., 2015) have improved ShedEx to incorporate the multi-channel scenario. The proposed scheme claims to have reduced the overall delay by 20% in scheduling in comparison to ShedEx. In (Jiming et al., 2015), the authors propose a regret matching based channel assignment algorithm (RMCA). The scheme aims to minimize the overhead induced by the use of multi-channels. The paper validated the results with simulations and practical implementation where an overall improvement in performance and reduction in complexity was observed. In de Moraes and Silva (2014), the authors proposed an analytical approach to model the multi-channel scenario. Numerical and simulation based evaluation was used to authenticate the model.

3.2.4 Priority Based MAC Schemes

In IWSNs, the use of TDMA and multi-channel schemes offer suitable solutions for the regulatory control systems, open loop control systems, whereas the CSMA/CA based schemes are more suitable for alerting and monitoring systems. However, to deal with the emergency/safety systems and supervisory control systems, suitable solutions are required to assist timely channel access without the added delay due to TDMA schemes or unreliable and lossy links due to CSMA/CA based channel access. To ensure reliable real time data delivery priority based communications opens up new avenues. Since the majority of processes within industry have different levels of priority, therefore it makes more sense to deal them accordingly. In priority aware MAC protocols, the priority of different communications traffic is established and is used as a precedence to decide which traffic/data is to be transmitted first. Priority based MAC schemes are proposed in the literature, but such work is relatively limited. Some of the priority based MAC protocols are included in (de Moraes and Silva 2014–Zheng et al. 2015). In (Andersson et al., 2008), the authors used message content to prioritize communications within the network. In this protocol, the communications are also prioritized based on time deadline; however, the assumptions in this paper are relatively unrealistic, such as full duplex communications being assumed. Another MAC protocol for critical processes is proposed in Meng et al. (2015). The scheme uses priority based communications where traffic is divided in four groups where the priority based protocol gives higher precedence to more critical traffic compared to less critical traffic. In Wei et al. (2014), the authors propose an arbitration frequency based priority enabled

MAC protocol. The protocols allocates arbitration frequency to the users and, based on the allocated arbitration frequency, the critical nature of communications is decided. The protocol is evaluated using the discrete time Markov chain model and guaranteed access of the highest priority user is assured.

While there exist some protocols which offer priority based communications yet the developments in priority based MAC protocols is relatively limited and most of the proposed schemes are static in nature. In this chapter, a dynamic TDMA based channel access scheme is presented that targets low latency communications of critical data within a timely and reliable fashion.

3.3 Priority Based Information Scheduling and Transmission

TDMA based communications are preferred in industry. However, to incorporate supervisory and emergency communications, an on demand channel access is required. In addition, due to the critical nature of emergency traffic, low latency is to be ensured from channel request to channel allocations.

This section presents a MAC based channel access scheme that takes into consideration critical information and schedules channel time slots to facilitate communication. Since emergency communications and supervisory control communications are asynchronous in nature, allocation of channels is made through requests on the control channel instead of allocating regular time slots in the superframe. The presented scheme in (Raza et al., 2018b) uses the superframe format of IEEE with some added changes to facilitate timely communications of emergency traffic. For communications, the IEEE MAC low latency deterministic network (LLDN) is used, which is updated to support emergency communications. The conventional superframe of IEEE 802.15.4e MAC LLDN is presented in Figure 3.2. As represented in the figure, each superframe is divided into n timeslots, each of duration t. TN1 represents the transmission of N1 (Node 1), G-Ack is the group acknowledgement, and S-Slot represents the shared slot. The updated superframe to incorporate emergency communications is presented in Figure 3.3.

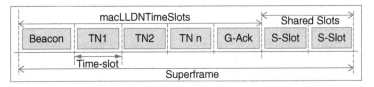

Figure 3.2 MAC LLDN superframe with communication and shared slots.

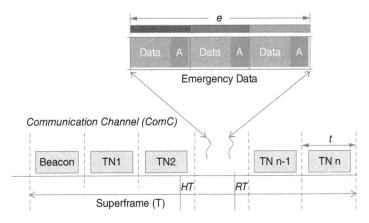

Figure 3.3 Updated superframe to incorporate emergency communications.

The communications in the presented system use star topology where the data is accumulated at the cluster-head. As depicted in Figure 3.3, the emergency nodes within the network can request an immediate channel access whenever a situation arises. The emergency data requests are made through a control channel where the regular information is stopped for a limited time to give immediate access to emergency data. The prescheduled communications are stopped using the HT sequence. Once the HT sequence is broadcast by the cluster all the communications from the regular scheduled nodes is stopped until the RT sequence is broadcastd by the cluster head. During the window where the regular communications stop, the emergency data is transmitted to the cluster head for immediate action. Since the communications from one sensor take place in a single time slot, ideally the regular communications are stopped for one time slot. However, multiple requests can be made at a particular time and so the time slots allocated to emergency communications will be dependent on the number of channel requests received.

The performance of the presented protocol is compared with the IEEE 802.15.4e, MAC LLDN framework, where a mathematical model for both is presented.

The superframe duration of LLDN is represented as T_{LLDN}, whereas in the case of the superframe duration of the proposed protocol T_p is given by

$$T_p = T_{\text{LLDN}} + e \tag{3.1}$$

where e is the added duration of emergency time slots in the superframe. Due to the asynchronous nature of the occurrence of an emergency event, emergency communication is an event driven in nature and therefore is modelled as a Poisson (α) distribution. The duration of timeslot (t) is presented in Equation 3.2 whereas the probability mass function of the emergency occurrences is presented in Equation 3.3.

$$t = \frac{T_{\text{LLDN}} - \text{MAC\_Payload}}{n}. \tag{3.2}$$

Here n is the number of time slots in the superframe whereas MAC_Payload is taken to be 3.84 ms.

$$P_X(x) = \begin{cases} \dfrac{\frac{\alpha^x e^{-\alpha}}{x!}}{\sum_{y=0}^m \frac{\alpha^y e^{-\alpha}}{y!}} & \text{where } x = 0, 1, 2, \ldots, m \\ 0 & \text{otherwise.} \end{cases} \tag{3.3}$$

Here m refers to the number of emergency nodes included in a cluster and $\alpha = \lambda t$, where λ represents the average occurrence of an emergency scenario. The total duration of added emergency slots is given by

$$e = t \times \left\{ \left(\sum_{x=0}^m \left(x \times \frac{\frac{\alpha^x e^{-\alpha}}{x!}}{\sum_{y=0}^m \frac{\alpha^y e^{-\alpha}}{y!}} \right) \right) \pm \left(\sum_{x=0}^m \left(x - \sum_{x=0}^m \left(x \times \frac{\frac{\alpha^x e^{-\alpha}}{x!}}{\sum_{y=0}^m \frac{\alpha^y e^{-\alpha}}{y!}} \right) \right) \right)^2 \right\}. \tag{3.4}$$

The average access delay of LLDNs (d_{LLDN}) and the average access delay of the proposed scheme (d_p) is presented in Equations 3.5 and 3.6 respectively. Whereas the average delay in successful communications in both LLDN and the proposed ($d_{\text{LLDN\_s}}$, d_{p_s}) are presented in Equations 3.7 and 3.8.

$$d_{\text{LLDN}} = \frac{1}{2} T_{\text{LLDN}} \tag{3.5}$$

$$d_p = \sum_{x=1}^{m} \left[\left(\delta t + \frac{1}{2}t + (x-1) \times t + \left(\frac{x}{n} \times \text{PL}_\text{d}\text{elay} \right) \right) \left(\frac{\alpha^x e^{-\alpha} / x!}{\sum_{y=1}^{m} \alpha^y e^{-\alpha} / y!} \right) \right] \tag{3.6}$$

$$d_{\text{LLDN\_s}} = d_{\text{LLDN}} \times \sum_{w=1}^{\infty} w \times p(1-p)^{w-1} \tag{3.7}$$

$$d_{p\_s} = d_{\text{EE-MAC}} \times \sum_{w=1}^{\infty} w \times p(1-p)^{w-1}. \tag{3.8}$$

Here p is the probability of successful transmission and w is the count of transmissions until the successful communication takesplace.

The performance of the proposed protocol is compared with IEEE802.15.4e LLDN, where an overall reduction in average access delay has been witnessed.

In Figure 3.4, the average access delay for MAC LLDN and the proposed MAC scheme is presented as a function of number of emergency

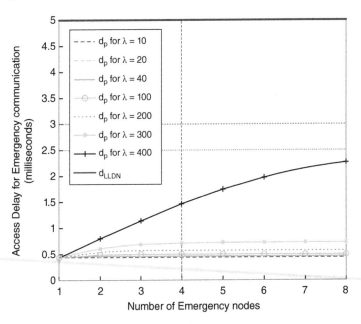

Figure 3.4 Channel access delay as a function of number of emergency nodes (m).

nodes. Since the LLDN offers a TDMA based access, the emergency nodes have to wait for the designated slot to access the channel, thus resulting in uniform average access delay, as represented in the figure. By contrast, the proposed scheme targets on demand channel access, therefore the access delay is relatively lower than that of LLDN. The figure shows that the EE-MAC even under extreme conditions [$m = 8$, $\lambda = 400$ emergency communication requests per second (on average)] manages to offer a 50% reduction in the access delay in emergency communications. In addition for less extreme cases the delay is maintained below 1 ms.

To further investigate the delay in emergency communication, the average delay till successful communication for the different values of λ (the number of emergency communication requests per second) and m (number of emergency nodes) is presented in Figure 3.5. The figure shows that the average successful communication delay in the proposed work, even under poor channel conditions ($p = 0.7$) and higher number of emergency requests (400 requests per second, on average),

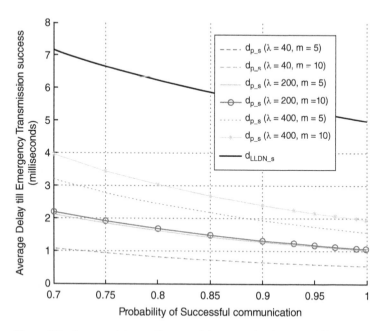

Figure 3.5 Average delay until successful communication with different channel conditions (p).

manages to reduce the delay to only 4 ms. For less critical channel and lower emergency channel requests, the overall delay is reduced up to 90% in comparison to IEEE 802.15.4e LLDN.

With the incorporation of a control channel based request mechanism and using reservation of slots for emergency communications once the request has been made allows improved low latency communications of the critical information to the control centre. The delay observed is within the acceptable range and offers relatively faster access to the channel in comparison to LLDN.

3.4 Summary

Delay is an important attribute in IWSNs. Any added delay affects the performance/efficiency of the implemented process control and system stability. In the proposed work, an improved MAC scheme is presented that allows the emergency nodes to requests channel access on a priority basis. This reduces the overall delay of channel access as well as successful communications, notably even within harsh circumstances, thus offering improved channel access delay compared to IEEE802.15.4e LLDN.

Bibliography

Anastasi G., Conti M. and Di Francesco M. (2011) A comprehensive analysis of the MAC unreliability problem in IEEE 802.15. 4 wireless sensor networks. *IEEE Transactions on Industrial Informatics*, 7, 52–65.

Andersson Björn, Pereira Nuno, Elmenreich Wilfried, Tovar Eduardo, Pacheco Filipe and Cruz Nuno (2008) A Scalable and Efficient Approach for Obtaining Measurements in CAN-Based Control Systems. *IEEE Transactions on Industrial Informatics*, 4, 80–91.

Bachir A., Dohler M., Watteyne T. and Leung K. K. (2010) MAC essentials for wireless sensor networks. *IEEE Communications Surveys & Tutorials*, 12, 222–248.

Chen A.-W., Shih B., Chang C. and Chen C. (2010) Enhanced MAC channel selection to improve performance of IEEE 802.15. 4.

International Journal of Innovative Computing, Information and Control, 6, 5511–5526.

Demirkol I., Ersoy C. and Alagoz F. (2006) MAC protocols for wireless sensor networks: a survey. *IEEE Communications Magazine*, 44, 115–121.

de Moraes L. F. and Silva R. S. (2014) Analysis of multichannel wireless networks With priority-based polling MAC protocols. *Wireless Days (WD)*, 2014 IFIP, 1–6.

Dobslaw F., Zhang T. and Gidlund M. (2014) End-to-End Reliability-aware Scheduling for Wireless Sensor Networks. *IEEE Transactions on Industrial Informatics*, 1-1.

Dobslaw F., Zhang T. and Gidlund M. (2015) Latency Improvement Strategies for Reliability-Aware Scheduling in Industrial Wireless Sensor Networks. *International Journal of Distributed Sensor Networks*, 2015.

Doudou M., Djenouri D. and Badache N. (2013) Survey on Latency Issues of Asynchronous MAC Protocols in Delay-Sensitive Wireless Sensor Networks. *IEEE Communications Surveys & Tutorials*, 15, 528–550.

Ergen S. C. and Varaiya P. (2010) TDMA scheduling algorithms for wireless sensor networks. *Wireless Networks*, 16, 985–997.

Francesco M. D., Anastasi G., Conti M., Das S. K. and Neri V. (2011) Reliability and energy-efficiency in IEEE 802.15. 4/ZigBee sensor networks: an adaptive and cross-layer approach. *IEEE Journal on Selected Areas in Communications*, 29, 1508–1524.

Gungor V. C. and Hancke G. P. (2009) Industrial Wireless Sensor Networks: Challenges, Design Principles, and Technical Approaches. *IEEE Transactions on Industrial Electronics* 56, 4258–4265.

IEEE (2006) IEEE Standard for Information Technology—Local and Metropolitan Area Networks—Specific Requirements—Part 15.4: Wireless Medium Access Control (MAC) and Physical Layer (PHY) Specifications for Low Rate Wireless Personal Area Networks (WPANs), IEEE Standard 802.15.4-2006, IEEE, London.

Jiming C., Qing Y., Bo C., Youxian S., Yanfei F. and Xuemin S. (2015) Dynamic Channel Assignment for Wireless Sensor Networks: A Regret Matching Based Approach. *IEEE Transactions on Parallel and Distributed Systems*, 26, 95–106.

Kan Y., Gidlund M., Akerberg J. and Bjorkman M. (2013) Reliable real-time routing protocol for industrial wireless sensor and actuator networkstextit2013 8th IEEE Conference on Industrial Electronics and Applications (ICIEA) 1895–1901.

Kredo K. and Mohapatra P. (2007) Medium access control in wireless sensor networks *Computer Networks*, 51, 961–994.

Kumar A., Zhao M., Wong K., Guan Y. L. and Chong P. H. J. (2018) A Comprehensive Study of IoT and WSN MAC Protocols: Research Issues, Challenges and Opportunities. *IEEE Access*, 6, 76228–76262. https:doi:10.1109/ACCESS.2018.2883391

Langendoen K. (2008) Medium access control in wireless sensor networks *Medium Access Control in Wireless Networks*, 2, 535–560.

Marchenko N., Andre T., Brandner G., Masood W. and C. Bettstetter (2014) An experimental study of selective cooperative relaying in industrial wireless sensor networks. *IEEE Transactions on Industrial Informatics*, 10, 1806–1816.

Meng Z., Junru L., Wei L. and Haibin Y. (2015) A priority-aware frequency domain polling MAC protocol for OFDMA-based networks in cyber-physical systems. *Journal of Automatica Sinica, IEEE/CAA*, 2, 412–421.

Pangun P., Di Marco P., Fischione C. and Johansson K. H. (2013) Modeling and Optimization of the IEEE 802.15.4 Protocol for Reliable and Timely Communications. *IEEE Transactions on Parallel and Distributed Systems*, 24, 550–564.

Pei H., Li X., Soltani S., Mutka M. W. and Ning X. (2013) The Evolution of MAC Protocols in Wireless Sensor Networks: A Survey. *IEEE Communications Surveys & Tutorials*, 15, 101–120.

Naik P. and Sivalingam K. M. (2004) A survey of MAC protocols for sensor networks. Wireless Sensor Networks, Springer, Berlin, 93–107.

Raza M. , Hussain S., Le-Minh H., Aslam N. (2017) Novel MAC layer proposal for Ultra-Reliable and Low-Latency Communication in Industrial Wireless Sensor Networks; A proposed scenario for 5G powered Networks. *ZTE Communications*, S1, 006, 1673–5188.

Raza M., Aslam N., Le-Minh H., Hussain S., Cao Y. and Khan N. M., (2018a), A Critical analysis of research potential, challenges, and future directives in industrial wireless sensor networks. *IEEE Communications Surveys & Tutorials*, 20, 1, 39–95. https://doi:10 .1109/COMST.2017.2759725

Raza M., Aslam N., Le-Minh H. and Hussain S. (2018b) A Novel MAC proposal for Critical and Emergency Communication in Industrial

Wireless Sensor Networks. *Elsevier Computers and Electrical Engineering*. https://doi.org/10.1016/j.compeleceng.2018.02.027

Roy A. and Sarma N. (2010) Energy saving in MAC layer of wireless sensor networks: a survey. National Workshop in Design and Analysis of Algorithm (NWDAA), Tezpur University, India.

Shen W., Zhang T., Gidlund M. and Dobslaw F. (2013) SAS-TDMA: A source aware scheduling algorithm for real-time communication in industrial wireless sensor networks. *Wireless networks*, 19, 1155–1170.

Shibo H., Jiming C., Peng C., Yu G., Tian H. and Youxian S. (2012) Maintaining Quality of Sensing with Actors in Wireless Sensor Networks. *IEEE Transactions on Parallel and Distributed Systems*, 23, 1657–1667.

Wei S., Tingting Z., Barac F. and Gidlund M. (2014) PriorityMAC: A Priority-Enhanced MAC Protocol for Critical Traffic in Industrial Wireless Sensor and Actuator Networks. *IEEE Transactions on Industrial Informatics*, 10, 824–835.

Yan M., Lam K.-Y., Han S., Chan E., Chen Q., Fan P. et al. (2014) Hypergraph-based data link layer scheduling for reliable packet delivery in wireless sensing and control networks with end-to-end delay constraints. *Information Sciences*, 278, 34–55.

Yoo S.-e., Chong P. K., Kim D., Doh Y., Pham M.-L., E Choi. et al. (2010) Guaranteeing real-time services for industrial wireless sensor networks with IEEE 802.15. 4. *IEEE Transactions on Industrial Electronics*, 57, 3868–3876.

Zhao J., Qin Y., Yang D. and Rao Y. (2014) A source aware scheduling algorithm for time-optimal convergecast. *International Journal of Distributed Sensor Networks*, 2014.

Zheng T., Gidlund M. and Akerberg J. (2015) WirArb: A New MAC Protocol for Time Critical Industrial Wireless Sensor Network Applications. *IEEE Sensors Journal*, 1-1.

4

Narrowband Internet of Things (NB-IoT) for Industrial Automation

Hassan Malik[1], Muhammad Mahtab Alam[1], Alar Kuusik[1], Yannick Le Moullec[1], and Sven Pärand[2]

[1] Thomas Johann Seebeck Department of Electronics, Tallinn University of Technology, Estonia
[2] Telia Estonia Ltd., Estonia

4.1 Introduction

This chapter presents an insight into the narrowband Internet of Things (NB-IoT), a cellular technology for enabling massive machine type communications such as automation in Industry 4.0. Section 4.2 of this chapter provides an overview of NB-IoT with respect to industrial automation. Section 4.3 highlights the design of NB-IoT to fulfil the intended objective, namely flexible deployment, extended coverage, long battery life, low device complexity, ultra-low device cost, small data transmission, and massive device connectivity. Section 4.4 provides a detailed overview of both downlink and uplink radio frames and transmission schemes along with the physical control channels. Section 4.5 includes the specific requirements of industrial automation use cases to highlight the feasibility of NB-IoT for such scenarios. Finally, Section 4.6 concludes the key features of NB-IoT that can act as a key enabler for industrial automation.

4.2 Overview of NB-IoT

Industry 4.0 refers to the digitalization of the vertical and horizontal value chain, automatization and integration by implementing various

Wireless Automation as an Enabler for the Next Industrial Revolution, First Edition.
Edited by Muhammad A. Imran, Sajjad Hussain and Qammer H. Abbasi.
© 2020 John Wiley & Sons Ltd. Published 2020 by John Wiley & Sons Ltd.

technological concepts such as cyber physical systems, the Internet of Things and smart factory, which enable communication between products, their environment and the business side are the key focus areas of today's world. The convergence of these technologies will enable seamless connectivity of industrial processes, assets, workforce and users to exchange information and make a proactive decision on an upcoming event to better manage production and increase profitability. One of the main characteristics of Industry 4.0 is its diverse applications. For example, health and safety monitoring of employees or factory production processes have to exchange a huge amount of data while being stationary, whereas devices for in-house fleet management, i.e. robots, have a small amount of data but require frequent handovers. Furthermore, there are other applications such as monitoring of critical asset location, the output of workstations, stock level, air pressure, electricity consumption, environment factors and the performance of production tools such as precision screwdrivers or condition monitoring. Most of these processes are stationary with a small amount of data and are delay insensitive. However, with such digitization in industry, wireless IoT devices are expected to grow exponentially and sustaining continued growth requires more spectrum or efficient ways of using the available spectrum. In this regard, NB-IoT is specifically designed to provide seamless connectivity to a large number of devices with a system bandwidth of 180 kHz. NB-IoT is designed to support a massive number of devices in a small area with extended coverage, long battery life, low complexity and low cost.

NB-IoT is a cellular based radio access technology introduced by the 3rd Generation Partnership Project (3GPP) in 2016 to address the fastest growing market of low-power wide area connectivity enabling massive machine type communication (mMTC) (3GPP TS 36.300, 2017). Since then, continuous efforts to improve the technology have resulted in Release 14 and Release 15 in 2017 and 2018, respectively (3GPP RP-161901, 2016; 3GPP RP-170852, 2017). Table 4.1 summarizes the key design features included in all the three NB-IoT releases.

NB-IoT is a part of the 3GPP Long Term Evolution (LTE) specifications and inherits most of the design from LTE, which has resulted in a simplification of the standardization process and ensures fast time to market. Due to the coherence with LTE design, it is also possible

Table 4.1 NB-IoT key design features (Ratasuk *et al.*, 2017).

3GPP release	Features
Release 13	180 kHz system bandwidth, control channel design modifications, support for half-duplex frequency division duplex (HD-FDD), deployment flexibility (i.e. in-band, standalone, guard band), single-tone and multi-tone transmission support in uplink, radio resource control (RRC) connection suspend/resume, control plane optimization for small data transmission, extension in discontinuous reception (eDRX) mechanism, mobility support in idle-mode.
Release 14	Mobility support, multi-cast group communication support, enhanced data rate, support for paging and random access in non-anchor carrier, introduction of new power class (14 dBm).
Release 15	Latency reduction, improved measurement accuracy, enhancement in random access procedure, support for time division duplex (TDD) mode.

to deploy NB-IoT deployed as a software upgrade in already existing networks. This will result in global market outreach and fast adaptation of the technology as there will be no new infrastructure costs. NB-IoT is designed for ultra low cost mMTC to provide long battery life and short packet transmissions (Wang *et al.*, 2017b). Furthermore, it is designed to provide enhanced coverage over the General Packet Radio Service (GPRS) to enable deep penetration in indoor and underground environments.

In order to achieve these design objectives, some of the basic LTE features have been excluded from NB-IoT, reflected by, e.g., the lack of LTE-WLAN interworking, interference avoidance for in-device coexistence and measurements to monitor the channel quality. Most of the LTE-Advance features are also not supported, such as carrier aggregation, device-to-device services, and dual connectivity. Moreover, there is no quality of service (QoS) concept; NB-IoT is designed for delay tolerant data transmission. Similarly, services that require guaranteed bit rates are not supported by NB-IoT. Therefore, user equipment that support and work on NB-IoT technology is categorized as LTE Cat-NB1.

4.3 NB-IoT Design Characteristics

In this section, we highlight the main design objectives of NB-IoT that meet the requirements of mMTC and the corresponding solutions adopted to achieve the design requirements.

4.3.1 Low Device Complexity and Low Cost

One of the key design requirements for NB-IoT is to provide low device complexity and low cost, which usually depends on the baseband processing, memory storage and radio frequency (RF) components. Therefore, NB-IoT is designed for low complexity receiver processing by reducing the system bandwidth to a narrow bandwidth, i.e. 180 kHz and allowing one synchronization sequence for both time and frequency synchronization to the network. The devices have the provision to use a low sampling rate of up to 240 kHz and use the synchronization properties for complexity reduction. Moreover, in terms of channel coding, NB-IoT supports traditional convolution codes, i.e. tail biting convolution codes (TBCC) in the uplink instead of LTE turbo code due to their iterative processing (Liberg *et al.*, 2018). Additionally, NB-IoT is also restricted to using low-order modulations (i.e. $\pi/2$ binary phase shift keying (BPSK) and $\pi/4$ quadrature phase shift keying (QPSK) in uplink and QPSK in downlink) and has no support for multiple-input multiple-output (MIMO) transmission.

For further reduction of processing and complexity, a restriction on the transport block size (TBS) in both uplink and downlink is applied with maximum TBS of 680 bits and 1000 bits, respectively, in Release 13. However, limiting the transport block size also limits the achievable data rate, i.e. 62.5 kbps in uplink and 25.5 kbps in downlink. No doubt, the target of NB-IoT is to achieve low complexity; however, these numbers seem to be on the lower end, especially for downlink, which is expected to support a large data transfer for firmware download to the device. Therefore, in Release 14, the maximum TBS is increased to 2536 bits in both downlink and uplink, due to which the achievable data rates are improved to 106 kbps for uplink and 79 kbps for downlink. The increase in TBS results in a slight increase of soft buffer memory requirements and computational complexity as more processing needs to be performed during the decoding phase of the data block.

For reducing the RF receiver chain complexity and cost, NB-IoT devices support single radio access technology with one transmit and receive antenna. This means neither downlink receiver diversity nor uplink transmit diversity is required in the device. Also, the device can only operate in half-duplex frequency division duplexing (HD-FDD) in Releases 13 and 14, which means it cannot listen to downlink while transmitting in uplink and vice versa. NB-IoT also allows relaxed oscillator accuracy, i.e. a device can achieve initial acquisition with an inaccuracy of up to 20 ppm. Furthermore, the data transmission scheme is devised to easily track the frequency offset. As NB-IoT is HD-FDD, a duplexer is not required in the RF chain of the device, which reduces the device cost. Although TDD support was introduced in Release 15, it is still in its development phase.

NB-IoT devices operate with a maximum transmit power of either 20 or 23 dBm in Release 13, which allows on-chip integration of the power amplifier resulting in cheap device manufacturing. However, the power amplifier's drain current in both of these power classes is expected to exceed 100 mA, which limits the use of small coin-cell batteries. To facilitate the use of such batteries for applications like smart fitness bands and watches, or for asset monitoring where large battery deployment is not possible, a new power class support of 14 dBm was introduced in Release 14. However, with the reduction of power, the uplink coverage will also be significantly reduced; nevertheless, the maximum coupling loss (MCL) for 14 dBm power class is 155 dB, which is still higher than the traditional GPRS and LTE systems by 10 dB.

4.3.2 Coverage Enhancement (CE)

Coverage enhancement is one of the key requirements for MTC as most of the application scenarios are either in an indoor or underground environment where the signal penetration loss is significantly higher as compared to an outdoor environment. In order to provide a reliable communication link, an extension of coverage with respect to traditional LTE or GPRS systems is of utmost significance. In NB-IoT, repetitions of both data and control packets are used as a base solution for coverage enhancement, achieving an extra coverage of 20 dB compared to GPRS with a power class of 23 dBm. All repetitions are self-decodable, and repetitions are acknowledged only once. Based on the MCL, NB-IoT specifies three coverage classes, i.e. 0 dB (normal), 10 dB (robust) and 20 dB (extreme), and the maximum number of repetition allowed in downlink and uplink are 2048 and 128, respectively (Chafii *et al.*, 2018). In NB-IoT, all channels can use repetition to enhance coverage. Besides the repetitions, NB-IoT design allows using a close-to-constant envelop waveform in uplink, which reduces the need to back off the output power from the maximum level. This helps to preserve the best coverage for a given power amplifier in extreme coverage and power limited situations.

4.3.3 Long Device Battery Lifetime

NB-IoT is designed to help IoT devices conserve battery power and potentially achieve a 10 year battery lifetime (Wang *et al.*, 2017). Device power consumption depends on how the device behaves in its idle time and how frequently the transmission takes place. In most IoT applications, transmission occurs on a set interval of time;

during the rest of the time, the device is in idle mode. In traditional LTE or GPRS systems, the device in idle mode needs to monitor paging and mobility measurements on a regular basis to remain connected to the network. A reduction of power consumption can be achieved by increasing the period between paging monitoring or simply allowing the device to completely shut down the monitoring of paging and other network measurements. In this regard, NB-IoT supports two types of schemes for power conservation, namely, power saving mode (PSM) and extended discontinuous reception (eDRX) (Martínez *et al.*, 2018).

The PSM feature was introduced in Release 12 and is also available for all LTE device categories. PSM is similar to power-off; however, the device still remains registered with the network and when the device wakes up, it does not have to perform the registration process with the network. This reduces the signalling overhead and optimizes the device power consumption. The device can request the PSM mode by setting the tracking area update (TAU) timer. The maximum time the device can be in PSM mode is approximately 413 days in Release 13 (3GPP TS-24.301 v13.8.0, 2016).

eDRX is an extension of an existing LTE feature introduced in 3GPP Release 13 that enables IoT devices to connect to the network on a need basis. Devices can remain inactive or in sleep mode for hours. The network and devices can negotiate the period of time for which the device can be in sleep mode, expressed in terms of the number of hyper frames (i.e. 1024 frames or 10.24 s); this can be up to 3 hours in NB-IoT. During this period, the device switches off its receiver and is not listening to paging or downlink control channels. After the expiry of the set period, the device wakes up and starts listening. Furthermore, in Release 13, NB-IoT also offers connected mode DRX (cDRX), which can be configured from 2.56 to 10.24 s. eDRX can be configured without PSM or with PSM to achieve maximum power savings.

4.3.4 Massive Device Support

The aim of NB-IoT is to support massive device connectivity. NB-IoT uplink operates on single carrier frequency division multiple access (SC-FDMA) with two different sub-carrier spacing configurations, i.e. 15 kHz and 3.75 kHz. The resource grid in uplink and downlink with

Table 4.2 NB-IoT uplink resource unit configuration.

Configuration	Sub-carrier spacing (kHz)	Required sub-carriers	Duration (ms)
1	15	1	8
2	15	3	4
3	15	6	2
4	15	12	1
5	3.75	1	32

a sub-carrier spacing of 15 kHz remains the same as in LTE, which allows seamless integration on the LTE carrier. However, with 3.75 kHz, the symbol duration increased to four times (i.e. 2 ms slot duration) as compared to 15 kHz and provides 48 sub-carriers in the frequency domain.

Furthermore, to provide massive device support, NB-IoT has the provision to share the uplink bandwidth with multiple devices in the frequency domain of the NB-IoT carrier rather than a complete resource block allocated to a single device, as is the case for LTE. Therefore, NB-IoT introduces the concept of a resource unit (RU) where the device can be allocated a different number of sub-carriers and the allocation can last for different durations, which is referred to as single-tone or multi-tone transmission. Multi-tone transmission is only possible with the 15 kHz sub-carrier spacing and the recommended group of sub-carriers allocated simultaneously are 3, 6 and 12 for a time duration of 4 ms, 2 ms and 1 ms, respectively (Wang *et al.*, 2017a). On the other hand, single-tone transmission allows the maximum amount of simultaneous transmission per NB-IoT carrier at the cost of a large delay, i.e. 8 ms and 32 ms for 15 kHz and 3.75 kHz sub-carrier spacing, respectively. Table 4.2 presents the set of RU configurations as specified in the standard. Configurations 1 and 5 represent single-tone transmission and configurations 2, 3 and 4 represent multi-tone transmission.

4.3.5 Deployment Flexibility

To support maximum flexibility of deployment and co-existence with the legacy LTE and GSM systems, NB-IoT supports three modes of operation, i.e. standalone, in-band and guard-band (Ratasuk *et al.*, 2016).

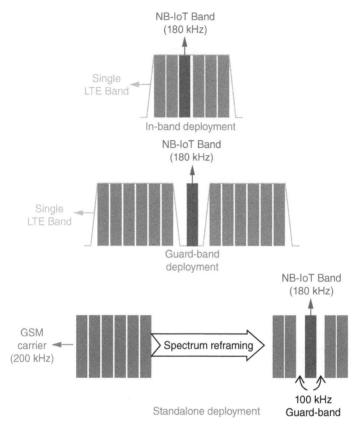

Figure 4.1 NB-IoT deployment modes.

Stand-alone deployment of NB-IoT can be done by using any available bandwidth larger than 180 kHz. This deployment strategy is particularly meant for GSM operators to deploy NB-IoT by reframing part of the GSM band, as shown in Figure 4.1, as NB-IoT needs to meet the GSM spectral mask, which is based on 200 kHz channelisation. For standalone deployment in the GSM system, an additional guard band is required between the GSM carrier and NB-IoT carrier. It is recommended that, in the case of the same operator, a guard-band of 100 kHz is needed whereas, in the case of different operators, a guard-band of 200 kHz should be left empty on each side of the NB-IoT carrier. NB-IoT is designed to be deployed in the existing LTE network, either by using one of the LTE physical resource block (PRB) or LTE guard-band referred to as in-band and guard-band deployment, respectively, as presented in Figure 4.1. In band-band deployment, the assignment of resources between LTE and NB-IoT is not fixed. Furthermore, there is a certain restriction on frequencies, i.e. the PRB within the LTE carrier to be used for cell connection, as illustrated in Table 4.3. As seen in the table, in-band deployment of NB-IoT is not supported in LTE with 1.4 MHz bandwidth. Furthermore, NB-IoT should take into account the resource used by the LTE system such as the cell specific reference signal (CRS) and downlink control channel while allocating the resource for NB-IoT. It is worth noting that the resource allocated for synchronization signals in LTE, i.e. six inner PRBs cannot be used for NB-IoT deployment. However, in the case of guard-band deployment, no such restriction is imposed and a complete PRB can be used for NB-IoT.

4.3.6   Small Data Packet Transmission Support

Most of the IoT applications require small data packets to be transmitted to the network. In this regard, to reduce the required signalling process before the transmission of data to support such small packets with reduced overhead and delay, NB-IoT provides two optimization procedures (Andres-Maldonado *et al.*, 2017):

1) Control plane cellular Internet of Things (CIoT) evolved packet system (EPS) optimization (CP)
2) User plane CIoT EPS optimization (UP)

Both of these optimizations improve the transmission of a small burst of data as compared to traditional LTE networks.

4.3.6.1   Control Plane CIoT EPS Optimization (CP)

Control plane optimization allows the transmission of encapsulated user data using a control plane signalling message as a non-access stratum protocol data unit (NAS-PDU). As no data plane setup (i.e. access stratum (AS) security setup and user plane bearers establishment) is required, this results in a reduction of control plane messages for small data transmissions. For NB-IoT devices, the support for CP is mandatory. Furthermore, when a device transmits using CP, it can include a release assistance information (RAI) field in the encapsulated packet to notify the network if no further transmission required in uplink or downlink, or only downlink transmission, is expected to this uplink

Table 4.3 Allowed LTE PRB indices for cell connection in NB-IoT in-band operation (J. Schlienz and D. Raddino, 2016).

LTE system bandwidth	3 MHz	5 MHz	10 MHz	15 MHz	20 MHz
Allowed LTE indices for NB-IoT anchor carrier	2, 12	2, 7, 17, 22	4, 9, 14, 19, 30, 35, 40, 45	2, 7, 12, 17, 22, 27, 32, 42, 47, 52, 57, 62, 67, 72	4, 9, 14, 19, 24, 29, 34, 39, 44, 55, 60, 65, 70, 75, 80, 85, 90, 95

transmission (i.e. acknowledgement packet). In this case, the network immediately triggers the S1 Release procedure.

4.3.6.2 User Plane CIoT EPS Optimization (UP)

User plane optimization is also referred to as the RRC resume procedure where the basic idea is to resume configurations established in a previous connection between the device and the network. This requires the initial configuration of radio bearers and the AS security procedure between the network and device. After this, UP enables two new procedures to suspend and resume the RRC connection: connection suspend and resume. The support of UP is optional for NB-IoT devices. When the device goes to the RRC idle state, the connection suspend procedure enables and retains the device context at the device, evolved node B (eNB), and mobility management entity (MME). Whenever the device wants to start a new data transmission, it can resume the connection by providing the resume ID to the network. UP optimization enables avoiding the AS security and radio bearer configuration for each data transfer as compared to traditional LTE data transmission.

4.3.7 Multicast Transmission Support

NB-IoT introduces multi-cast downlink transmission based on single cell point to multi-point (SC-PTM) to provide support for efficient software upgrade and group message delivery services in Release 14 (Tsoukaneri *et al.*, 2018). For this, two logical channels, namely the single cell multicast control channel (SC-MCCH) and the single cell multi-cast traffic channel (SC-MTCH), are defined. Currently, in Releases 14 and 15, repetition is not supported for multi-cast transmission.

4.3.8 Mobility Support

NB-IoT provides positioning support in Release 14, which includes observed time difference of arrival (OTDOA) positioning, cell ID (CID) positioning, and enhanced cell ID (E-CID) positioning (Fischer, 2014).

OTDOA positioning is a downlink based positioning method in which the user device measures the time of arrival (ToA) of the reference signal received from the multiple transmission points and then reports the reference signal time difference (RSTD) to the location server for positioning purposes. On the other hand, the CID method estimates the position of the user device using geographical coordinates based on its serving cell. E-CID is an extension of CID in which the positioning is further improved by incorporating additional information such as narrowband reference signal received power (NRSRP), narrowband reference signal received quality (NRSRQ) and timing advance of the NB-IoT user device.

4.4 NB-IoT Frame Structure

Like in LTE, the frame structure of NB-IoT (shown in Figures 4.2 and 4.3) constitute a hyperframe cycle where each hyperframe cycle consists of 1024 hyperframes. Each hyperframe is constituted of

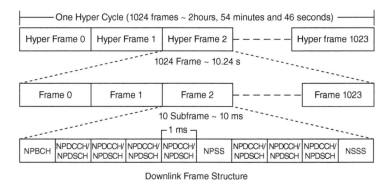

Figure 4.2 NB-IoT downlink frame structure.

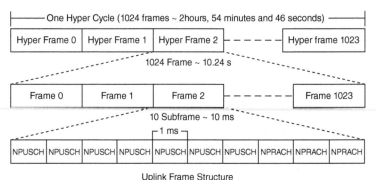

Figure 4.3 NB-IoT uplink frame structure.

1024 frames. Each frame consists of 10 sub-frames of 1 ms, and each sub-frame is sub-divided into two slots of 0.5 ms. In the frequency domain, in downlink, NB-IoT supports 15 kHz sub-carrier spacing which results in 12 sub-carriers and 7 symbols in each slot. One sub-carrier × one symbol constitutes one resource element (RE), the smallest transmission unit. In the uplink, along with 15 kHz sub-carrier spacing, NB-IoT provides additional support for 3.75 kHz sub-carrier spacing, which results in 48 sub-carriers and 7 symbols; however, the slot duration is increased, and the frame is directly divided into 5 slots of 2 ms each. The detailed description of downlink and uplink transmission with focus on control and signalling channels are presented below.

4.4.1 Downlink Transmission Scheme

NB-IoT employs orthogonal frequency division multiple access (OFDMA) in the downlink as in LTE. The slot duration, sub-carrier spacing, cyclic prefix (CP) duration are all identical as specified in LTE. The detailed frame structure of the NB-IoT downlink is illustrated in Figure 4.2 along with physical channels and signals. NB-IoT has two physical signals and three physical channels, as described in further details in Sections 4.4.1.1 to 4.4.1.5:

4.4.1.1 Narrowband Reference Signal (NRS)
A NRS is similar to a cell specific reference signal (CRS) in LTE, which is used for cell search and initial acquisition, measuring downlink channel quality, and downlink channel estimation for the coherent demodulation at the user device. Similar to LTE, NRS is transmitted in all sub-frames (i.e. broadcast or downlink transmission) occupying 8 REs for each antenna port. No doubt, the sequence for NRS is generated in a similar fashion as CRS in LTE; however, the mapping on REs is modified to avoid overlap with CRS, specifically in in-band deployment.

4.4.1.2 Narrowband Primary and Secondary Synchronization Signals (NPSS and NSSS)
A synchronization signal is used by user devices to synchronize downlink of NB-IoT cell in both time and frequency domains (i.e. NPSS) and it also enables devices to detect cell identification numbers (NB-IoT

PCI) along with frame structure information (i.e. NSSS). In NB-IoT, NPSS and NSSS are designed so that synchronization can be achieved during the initial acquisition phase without knowing the deployment mode of NB-IoT. However, to avoid collision with LTE, NPSS is transmitted in every fifth sub-frame with a periodicity interval of 10 ms and NSSS in every ninth sub-frame with an interval of 20 ms (i.e. 1 NSSS in every 2 frames). The remaining sub-frames (i.e. 1, 2, 3, 4, 6, 7, 8) are used by LTE.

4.4.1.3 Narrowband Physical Broadcast Channel (NPBCH)

NPBCH is similar to PBCH in LTE and carries a master information block (MIB), which includes initial cell access information such as system bandwidth, physical hybrid ARQ indicator channel structure, and the most significant eight bits of the system frame number; it is made up of 34 bits of information. However, after CRC and TBCC decoding it ends up with 1600 bits at the physical layer. Using QPSK modulation, it requires 800 REs to transmit the MIB completely. Therefore, 8 sub-frames are required to transmit the entire MIB and is transmitted in every 0th sub-frame (except for the initial three symbols of the sub-frame as they are used by LTE). One MIB transmission remains unchanged for 640 ms, after which it is re-initialized.

4.4.1.4 Narrowband Physical Downlink Control Channel (NPDCCH)

A NPDCCH carries a downlink control indicator (DCI), which is the logical block that includes the information related to downlink and uplink scheduling, random access channel (RACH) initiation and paging information. Unlike LTE, NB-IoT has only three DCI formats, as shown in Table 4.4. Besides this, NPDCCH is also divided

Table 4.4 DCI formats defined for NB-IoT (J. Schlienz and D. Raddino, 2016).

DCI format	Size (bits)	Information included
N0	23	Uplink scheduling grant
N1	23	Downlink scheduling grant, RACH initiated with NPDCCH
N2	15	Paging or system information update

into two sub-blocks consisting of six consecutive sub-carriers called narrowband control channel elements (NCCEs). There are two types of NPDCCH, format 0 and format 1. NPDCCH format 0 carries only one NCCE and format 1 carries two NCCEs in the same sub-frame for robust transmission.

The possible location of NPDCCH in the frame is called the search space; NB-IoT has three possible types: type 1 for paging, type 2 for the random access procedure and type 3 for downlink and uplink information. A user device uses a radio network temporary identifier (RNTI) specific to each type to identify whether NPDCCH carries any data for it or not. For user specific search space, NPDCCH has a periodicity that varies from 4 ms to 2.2 min. Furthermore, NPDCCH coverage extension is achieved through repetition; up to 2048 repetitions are defined in NB-IoT.

4.4.1.5 Narrowband Physical Downlink Shared Channel (NPDSCH)

NPDSCH consists of user downlink data, control information and a narrowband system information block (NB-SIB). Information about acquiring NB-SIB1 is presented in NPBCH and all the information for other NB-SIBs are included in NB-SIB1. NB-SIB1 has a fixed periodicity of 2560 ms and is present in the fourth sub-frame and transmitted continuously for 16 sub-frames. NPDSCH has also the provision for repetition for up to 2048 times.

4.4.2 Uplink Transmission Scheme

NB-IoT uplink transmission is based on SC-FDMA as in LTE. As mentioned, NB-IoT supports both single-tone and multi-tone transmission in uplink with sub-carrier spacing of 15 kHz and 3.75 kHz. Besides this, the NB-IoT uplink consists of one physical signal and two physical channels, as illustrated in Figure 4.3 (Malik *et al.*, 2018).

4.4.2.1 Demodulation Reference Signal (DMRS)

DMRS is used for uplink channel estimation accuracy improvement and impact uplink reliability and throughput. DMRSs are transmitted in the uplink data transmission and multiplexed with the data. DMRSs are modulated in a similar way to the associated data. Based on the NPUSCH format, DMRS can be sent either in one or three symbols per slot.

4.4.2.2 Narrowband Physical Random Access Channel (NPRACH)

To maintain orthogonality among users, a timing advance command is sent by the base station to the users. That requires user device uplink timing, which is estimated using NPRACH. NPRACH resources occupy a consecutive set of resources, either 12, 24, 36 or 48 sub-carriers, and are located on a discrete set of sub-carrier range. Therefore, in Figure 4.3 contiguous sets of resources have been allocated to NPRACH for demonstration purposes; however, the position in the frame can be different. To provide coverage extension to the NPRACH, repetition of up to 128 times is defined in NB-IoT. The detailed description of the NPRACH structure can be found in (Lin *et al.*, 2016).

4.4.2.3 Narrowband Uplink Shared Channel (NPUSCH)

NPUSCH is the main uplink data carrying channel. However, unlike LTE, NPSUCH in NB-IoT can also be used to send control information. In this regard, two NPSUCH formats have been defined in NB-IoT. NPUSCH format 1 is defined for uplink data and NPUSCH format 2 is used for signalling hybrid automatic repeat requests (HARQ). For both formats, the coverage extension is provided by repetition and up to 128 times are allowed.

4.4.3 NB-IoT Design Modification in Relation to LTE

The control channel design of NB-IoT is slightly different than traditional LTE; these design changes are primarily motivated due to the limited bandwidth (i.e. 180 kHz) of NB-IoT, as most of the LTE control channels are designed to span over multiple PRBs. Table 4.5 presents the modifications in the NB-IoT control channel as compared to LTE. These changes ensure maintaining the NB-IoT frame structure and the best co-existence performance with the existing LTE system.

4.5 NB-IoT as an Enabler for Industry 4.0

The main advantages of Industry 4.0 use cases are resource efficiency, cost efficiency, worker support, QoS or production improvement, and seamless logistic support. NB-IoT as an enabler for low cost, low power and long-range communication can provide connectivity

Table 4.5 NB-IoT design modification in relation to LTE.

	Physical channel	NB-IoT design	LTE design
Downlink	NPSS	• New sequence to fit into one sub-frame • All cells are PSS	• LTE PSS is spanned over six PRBs • LTE uses three PSSs
	NSSS	• New sequence to fit into one sub-frame • NSSS provide system frame number in last three significant bits	• LTE SSS is spanned over six PRBs • LTE SSS has no such information.
	NPBCH	640 ms TTI	40 ms TTI
	NPDCCH	Can use multiple sub-frames in time	Use multiple sub-frames in frequency and one sub-frame in time domain
	NPDSCH	• Uses turbo code with one redundancy version • Uses maximum QPSK • Maximum TBS support up to 2536 bits • Single layer transmission support	• Uses turbo code with multiple redundancy version • LTE can use higher order modulation • TBS supports up to 70000 bits (without spatial multiplexing) • Supports multiple spatial-multiplexing layer transmission
Uplink	NPRACH	New preamble format based on single tone transmission with 3.75 kHz subcarrier spacing	LTE PRACH uses six PRBs and multi-tone transmission using 1.25 kHz sub-carrier spacing
	NPUSCH	• Bandwidth allocation smaller than PRB • Supports 15 kHz and 3.75 kHz numerology • Maximum TBS support up to 2536 bits • Uses $\pi/2$ BPSK or $\pi/4$ QPSK • Single layer transmission support	• Minimum bandwidth allocation is of 1 PRB • Supports only 15 kHz numerlogy • TBS supports up to 70000 bits • Supports higher order modulations • Supports multiple spatial-multiplexing layer transmission

to large number of devices and is able to meet the requirements of various industrial automation use cases that do not have stringent latency requirements. This section presents some of the use cases that can be optimized with NB-IoT. In this regard, Industry 4.0 has been grouped into four major categories of use cases, namely process automation, human–machine interfaces, logistics and warehousing, and monitoring and maintenance (Zhong *et al.*, 2017). In all these broader categories, there are various use case scenarios which require massive device connectivity with relaxed latency constraints for which NB-IoT is best suited. Some of these are as follows.

4.5.1 Process Automation

Process automation is referred to as the automation of industry processes such as production or assembly lines that are normally associated with continuous operation with specific requirements of determinism, reliability, functional safety and cyber security. Process automation use cases cover a large industrial landscape such as oil and gas, mining, refining, chemical plants, pharmaceutical, water management, steel, etc. This diversity means that process automation requires a wide communication coverage.

For example, in the case of mining or construction companies, sites are crowded with vehicles and machines such as trucks, wheel loader, drillers or even robots to perform a variety of tasks. In such high-risk environments, the ability to move assets from one place to another is key. Using driverless equipment in this situation is a potential time saver. For example, immediately after a blast, when people are not allowed in the area until the fumes have cleared, such autonomous equipment is still able to continue the job. However, in such applications, the latency and reliability requirement are not strict and NB-IoT is considered to be one of the key enablers for such use cases.

Another example, recently presented by Ericsson, is a use case for calibrating a high-precision screwdriver using NB-IoT in one of its factory as a pilot project Corporation (2018). The screwdriver normally requires routine calibration and lubrication depending on the utilization time. Enabling such smart solutions will not only result in time saving or labour reduction, it will also increase the quality of service for the end customer.

4.5.2 Human–Machine Interfaces

A human–machine interface comprises hardware and software solutions for information exchange among human operators and machines. It includes control and management of machine processes and can be performed with a simple touch input or a control panel for a complex industrial automation system. For example, using interactive user interfaces it is possible to observe the step-by-step processes of machines and to also provide instructions as needed. Furthermore, it can also enable a machine operator to call the technical support team instantly when the problem arises in the machine. The machine operator is also able to receive physical feedback using haptic technologies such as wearables non-obtrusively. New level safety can be achieved with such user interfaces, which is even more important than efficiency.

Furthermore, maintenance staff can also be able to visualize machine status and thus manage their workload accordingly. Remote collaboration with such human–machine interfaces will also require an off-site specialist to consult and guide local professionals. Similarly, managers can obtain the status of operations within the factory and can receive immediate notification of any potential risk yet to arise. Most of these applications do not have stringent latency restrictions and can be enabled with NB-IoT.

4.5.3 Logistics and Warehousing

Logistic and warehouse automation is another major use case for Industry automation 4.0 that can be enabled with NB-IoT. In this regard, asset tracking is one of the major applications in which it is possible to track goods once they leave the warehouse. With such an application, it is possible to monitor asset status, packages and even people (i.e. health and safety) in the supply chain. It also allows managing the production based on market demand and identifying opportunities for business improvement. Furthermore, within the warehouse or factory, tracking of critical equipment is also possible with such a solution. This will help to reduce the paperwork and the risk of equipment going missing and can be accessed quickly. Another use case scenario for NB-IoT is in fleet management solutions. Fleet operators are keen to reduce their total cost of ownership (TCO) while complying with strict environment and safety regulations. NB-IoT enabled applications can help fleet management by reducing fuel consumption, route management, driver performance monitoring and vehicle maintenance, etc.

4.5.4 Maintenance and Monitoring

Maintenance and monitoring is another category in Industry 4.0 where NB-IoT can play an important role. Sensors can be used to monitor a number of processes and conditions, particularly in a factory environment, such as temperature, humidity, noise level, different chemical emissions as well as production line equipment and production status, based on which a number of solutions can be provided. Some of these potential solutions can be warnings for the workforce in high-risk environmental conditions to ensure their safety, predictive maintenance of machines that can result in reduction of maintenance cost and potential delays, etc. Recently, Ericsson presented two use cases based on NB-IoT as part of their factory automation. One of the use cases is for monitoring environment parameters based on temperature and humidity level sensors to ensure the safety of staff; these environment parameters can also be used for some specific machines that need certain temperatures to operate. Another use case is based on pressure sensors to weigh storage boxes to monitor how full they are. Previously, such processes required manual staff to weigh the boxes and then record the information by scanning barcodes. As a whole, NB-IoT can be used for a range of use cases within the industrial automation process, ranging from monitoring of employee health and safety conditions to machine maintenance, production line flow control and even to asset tracking and management of the complete supply chain process.

4.6 Summary

Industrial automation is adopting IoT very rapidly. Mapping IoT applications in industrial automation with the recent development in communication technologies is essential for exploring future research directions and challenges. As a whole, it can be concluded that with the emergence of NB-IoT it is possible to meet the requirements of industrial automation applications that require low power, low cost, long-range data communication with a non-stringent delay budget. NB-IoT, thanks to its extended coverage of 20 dB as compared to traditional LTE and GPRS networks, can provide excellent connectivity within a factory environment where most deployments are either indoors or underground with significantly large penetration losses. Therefore, NB-IoT is a clear enabler for future smart factories.

Bibliography

3GPP RP-161901 (2016) New work item proposal: Enhancements of NB-IoT,RAN#73, *Techn. Ber.*.

3GPP RP-170852 (2017) New WID on Further NB-IoT enhancements,RAN#75,, *Techn. Ber.*.

3GPP TS-24.301 v13.8.0 (2016) Technical Specification Group Core Network and Terminals; Non-access-Stratum (NAS) Protocol for Evolved Packet System (EPS),, *Techn. Ber.*.

3GPP TS 36.300 (2017) Evolved Universal Terrestrial Radio Access (EUTRA) and Evolved Universal Terrestrial Radio Access Network (EUTRAN); Overall description, V13.7.0,, *Techn. Ber.*.

Andres-Maldonado, P., Ameigeiras, P., Prados-Garzon, J., Navarro-Ortiz, J., and Lopez-Soler, J.M. (2017) Narrowband IoT Data Transmission Procedures for Massive Machine-Type Communications. *IEEE Network*, 31 (6), 8–15.

Chafii, M., Bader, F., and Palicot, J. (2018) Enhancing coverage in narrow band-iot using machine learning, in *2018 IEEE Wireless Communications and Networking Conference (WCNC)*, S. 1–6.

Corporation, G. (2018) Industrial Iot Case study: Ericsson smart Factory, *Techn. Ber.*.

Fischer, S. (2014) Observed time difference of arrival (OTDOA) positioning in 3GPP LTE, Qualcomm Whitepaper.

J. Schlienz and D. Raddino (2016) Narrowband internet of things, *Techn. Ber.*.

Liberg, O., Sundberg, M., Wang, Y.P.E., Bergman, J., and Sachs, J. (2018) Chapter 7 - NB-IoT, in *Cellular Internet of Things*, Academic Press, S. 217–296.

Lin, X., Adhikary, A., and Wang, Y..E. (2016) Random Access Preamble Design and Detection for 3GPP Narrowband IoT Systems. *IEEE Wireless Communications Letters*, **5** (6), 640–643.

Malik, H., Pervaiz, H., Alam, M.M., Le Moullec, Y., Kuusik, A., and Imran, M.A. (2018) Radio resource management scheme in nb-iot systems. *IEEE Access*, **6**, 15 051–15 064.

Martínez, B., Adelantado, F., Bartoli, A., and Vilajosana, X. (2018) Exploring the Performance Boundaries of NB-IoT. *CoRR*, **abs/1810.00847**.

Ratasuk, R., Mangalvedhe, N., Xiong, Z., Robert, M., and Bhatoolaul, D. (2017) Enhancements of Narrowband IoT in 3GPP Rel-14 and Rel-15, in *2017 IEEE Conference on Standards for Communications and Networking (CSCN)*, S. 60–65.

Ratasuk, R., Vejlgaard, B., Mangalvedhe, N., and Ghosh, A. (2016) NB-IoT system for M2M communication, in *IEEE Wireless Communications and Networking Conference Workshops (WCNCW)*, S. 428–432.

Tsoukaneri, G., Condoluci, M., Mahmoodi, T., Dohler, M., and Marina, M.K. (2018) Group communications in narrowband-iot: Architecture, procedures, and evaluation. *IEEE Internet of Things Journal*, **5** (3), 1539–1549.

Wang, Y..E., Lin, X., Adhikary, A., Grovlen, A., Sui, Y., Blankenship, Y., Bergman, J., and Razaghi, H.S. (2017) A Primer on 3GPP Narrowband Internet of Things. *IEEE Communications Magazine*, **55** (3), 117–123.

Wang, Y..E., Lin, X., Adhikary, A., Grovlen, A., Sui, Y., Blankenship, Y., Bergman, J., and Razaghi, H.S. (2017a) A Primer on 3GPP Narrowband Internet of Things. *IEEE Communications Magazine*, **55** (3), 117–123, doi:10.1109/MCOM.2017.1600510CM.

Wang, Y.E., Lin, X., Adhikary, A., Grovlen, A., Sui, Y., Blankenship, Y., Bergman, J., and Razaghi, H.S. (2017b) A primer on 3gpp narrowband internet of things. *IEEE Communications Magazine*, **55** (3), 117–123.

Zhong, R.Y., Xu, X., Klotz, E., and Newman, S.T. (2017) Intelligent Manufacturing in the Context of Industry 4.0: A Review. *Engineering*, **3** (5), 616 – 630.

5

Ultra Reliable Low Latency Communications as an Enabler For Industry Automation

João Pedro Battistella Nadas[1], Guodong Zhao[1], Richard Demo Souza[2], and Muhammad A. Imran[1]

[1] *James Watt School of Engineering, University of Glasgow, UK*
[2] *Universidade Federal de Santa Catarina-Florianópolis, Brazil*

5.1   Introduction

The fifth generation of cellular networks, 5G, is set to enable an unprecedented number of use cases across all aspects of human activity. This is due to ambitious goals agreed upon by telecommunications experts aiming to provide unparalleled support for several *verticals*. These so-called verticals comprise the industries that leverage novel modes of operation envisioned as part of 5G. However, one might ask: why have these verticals not been supported by current Long Term Evolution (LTE) networks? The answer to this question is because, so far, cellular networks have been designed with one single purpose in mind: to serve mobile phone users.

This meant that across the entire network stack protocols and processes have been thought to improve key performance indicators (KPIs) and to provide the required quality of service (QoS) of the mobile phone use case. Moreover, with growing health and environmental concerns, future networks must be able to support all envisioned use cases while also being more energy efficient to guarantee sustainable growth. Thus, in order to enable the 5G vision of sustainable and ubiquitous connectivity, the technology currently used in LTE has to be redesigned from scratch, considering the QoS requirements of the several verticals. This means that 5G networks

Wireless Automation as an Enabler for the Next Industrial Revolution, First Edition.
Edited by Muhammad A. Imran, Sajjad Hussain and Qammer H. Abbasi.

will have to offer completely different capabilities in order to provide the required QoS.

One example of capability that will enable several of the novel use cases is ultra-reliable and low-latency communication (URLLC) 3GPP (2017), which consist of communication with unseen levels of reliability and with *guaranteed* maximum end-to-end latency. This is particularly challenging as, typically, these two requirements are antagonistic, since a faster transmission implies less reliable communication.

URLLC is the most important service required to enable true wireless factory automation, as mission-critical industrial processes can only switch to wireless technology if URLLC QoS can be attained. Typically, the requirements of wireless industrial automated communications fall below 10^{-9} in terms of reliability (i.e. error rate), whilst sub-millisecond latencies are needed at the same time Ashraf et al. (2017).

In other words, true wireless factory automation can only exist if there is support for URLLC. The motivation to add wireless capability in any industrial process is clear and can be summarized into three main reasons Chen et al. (2018). As it does not require deployment of cables and the proper structures to house the cabling, it is much cheaper to deploy and can greatly reduce the cost of installation of new equipment. Moreover, it is less subject to degradation over time, in particular considering an industrial process where moving parts are involved. Lastly, since the structures do not have to be fiscally connected, it offers new degrees of freedom and therefore enables new processes to be designed leveraging the added flexibility. In summary, wireless connectivity is more adaptable, cheaper to install and easier to maintain when compared to its wired counterpart Chen et al. (2018).

In this chapter, we are going to explore some of the technologies that are being developed today to enable URLLC. Moreover, we discuss industry automation applications that require URLLC and present existing wireless network solutions for industrial automation. Thus, the remainder of the chapter is organized as follows: Section 5.2 presents opportunities for URLLC in industry automation. Section 5.3 discusses existing wireless solutions. Section 5.4 presents enabling technologies developed to enable URLLC and Section 5.5 concludes the chapter.

5.2 Opportunities for URLLC in Industry Automation

In this section we discuss the applications and business models for URLLC in industry automation. Factory automation is a vague term and encompasses a wide range of applications. Some of these applications do not require URLLC capabilities, such as several process automation use cases, related to metallurgy, oil and gas, or simple chemical processes Luvisotto et al. (2017). Building automation to control lighting, temperature and other related factors also does not require low latency or high reliability and thus can be served with existing wireless technologies. On the other hand, other types of applications are much more demanding and require URLLC to be made wireless.

5.2.1 URLLC Industrial Applications

One class of applications that requires URLLC in order to operate are those with high-end motion control involved Orfanus et al. (2013). Examples of these applications include wind power systems, machine vision, servo-enabled mechanisms, etc Wang et al. (2010). Another example is found in power system automation such as fault detection in power distribution systems Popovski et al. (2013) or real-time detection and isolation of anomalies Feliciano et al. (2014). A third type consists of power electronics control, wherein the aim is to operate and control electronic circuits used in high-power applications Cottet et al. (2015). Control loops are of extreme importance in these applications and having one of the links in a control loop controlled via a wireless network is what is known as a wireless networked control system.

There are several topologies regarding wireless networked control systems. Next, we present three examples of topologies. Keep in mind that resources, such as the wireless channel, may be shared by several different plants in a factory. In Figure 5.1 we show a control loop where only the link between the plant and the controller is wireless. These types of systems consist of wireless connectivity between the sensors, attached to the plant, and the controller. They are useful in situations where deploying the sensors might be a difficult task or

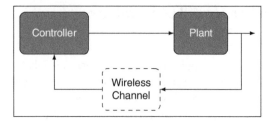

Figure 5.1 Example of a control loop topology where only the link between the plant and the controller is wireless.

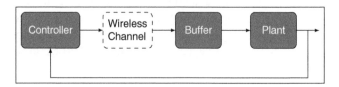

Figure 5.2 Example of a control loop topology where only the link between the controller and the plant is wireless.

where mobility is particularly important, making the use of cables less attractive. Moreover, in this topology, the controller can be attached to the plant to simplify the need for cabling between the controller and the actuators, and therefore one controller per plant is typically used. Since these sensors are typically operated by battery Cui et al. (2005) and communicate often, energy efficiency is one of the most important metrics in this case.

In Figure 5.2, on the other hand, we show an example where only the link between the controller and the plant is wireless. This can be used in the case where deploying a cable between the sensors and the controller is feasible but the actuators are positioned in difficult locations, or in places where mobility is required. As an example, we can think of a control loop where the sensor used for feedback is a camera that detects the position of a robotic arm, and this sensor is attached to a controller. Conversely, the actuators are attached to the robotic arm and therefore benefit greatly from having wireless connectivity. As, typically, the controller is attached to a power source, energy efficiency is not as relevant in these scenarios. Therefore, techniques that focus on improving the spectral efficiency are more relevant in this case. Also note that a buffer is represented collocated near the plant. This buffer is used when a model predictive controller (MPC) is employed

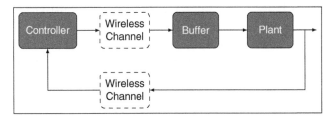

Figure 5.3 Example of a control loop topology where both links are wireless.

and more than one control signal is generated and transmitted. This provides additional reliability to the system as when packets fail a control signal from the buffer can be used in what is known as packetised predictive control (PPC)

Lastly, a combination of the two topologies exists, as depicted in Figure 5.3. Examples of applications consist of completely independent systems, where sensors, controller and the actuators are all located at different places in the plant. This becomes even more interesting when there is high mobility of the components. From a wireless engineering perspective, this is the most challenging topology, as the reliability of the loop depends on the reliability of both links. Moreover, as mentioned before, the sensor nodes are required to be energy efficient to increase the network lifetime.

5.2.2 New Business Models

The traditional cellular networks business model, wherein a small group of big telecommunications companies owns the infrastructure and relevant spectrum licenses, is inefficient as it considers a wide variety of services with diverse requirements. In 5G, on the other hand, the network will be jointly operated by several companies – both private and public – each owning a small part of the communication stack and/or spectrum and each providing specialized service depending on which application they want to support Elayoubi et al. (2017).

This enables very interesting business models that can be applied to factory automation. In particular, large industries are going to have the possibility of owning part of the communication infrastructure they are using in order to have more security in their process. At the same time, smaller industries could rent the service from other stake-holders, which will develop solutions for them, all in a unified 5G vision Elayoubi et al. (2017). In other words, the modular nature of 5G is one of the main enablers of its use in industrial automation and novel business models are certainly going to arise from this flexible structure, resulting in wealth generation.

5.3 Existing Solutions

5.3.1 LTE

LTE is the current technology deployed in cellular networks and its performance in terms of latency is of the order of 50 ms Larson et al. (2015) with a block error rate (BLER) performance of 10% before retransmissions, which suits the mobile broadband it was designed for but is nowhere near the requirements of the most stringent foreseen use cases, such as intelligent transportation, telesurgery and industry automation Chen et al. (2018).

Another issue with LTE networks is that, despite its nominal 50 ms latency, there is absolutely no guarantee that the communication will not be delayed further than that and typically the latency can be of the order of a couple of seconds Larson et al. (2015). An analysis of sources of delay in LTE networks has been presented by a 3rd Generation Partnership Project (3GPP) study 3GPP (2016) and the minimum delays that can be expected when using LTE are categorized below:

- Grant acquisition delay: occurs due to the grant based nature of LTE, in which the user has to request a resource block before initiating the transmission. This is done by emitting a scheduling request (SR), which can only occur at valid instants at the physical link control channel and thus contributes with 5 ms of added latency Chen et al. (2018).
- Random access delays: are introduced when the user is not connected to the base station and has to request a grant over a random access channel. Because it requires a handshake like protocol, it adds 9.5 ms to the user experienced delay Chen et al. (2018).
- Transmit time interval (TTI): is the time on air to actually send the messages. Because LTE uses orthogonal frequency division multiplexing (OFDM) the minimum value of TTI is the size of a

sub-frame chunk, which is 1 ms in current specifications 3GPP (2016).

- Signal processing delays: are largely dominated by the complexity of currently used algorithms. They are of the order of 4 ms for the currently used turbo coding Shirvanimoghaddam et al..
- Retransmission delay: is caused because of the scheduling structure of LTE. When a packet delivery fails the receiver has to wait 4 ms to send the non-acknowledgment (NACK) and then the transmitter has to wait an additional 4 ms to send a new attempt. Thus, a hybrid automatic repeat request adds 8 ms of latency per failed round 3GPP (2016).
- Core network delays: occur when the network is congested and packets are queued at some intermediate node and can vary greatly Chen et al. (2018) from almost no delay to a few extra seconds.

As we can see, current LTE technology cannot be used for URLLC wireless factory automation, as the required latency is on the order of 1 ms while the minimum we can reasonably expect when considering LTE is 6 ms if the node is already aligned with the base station; no retransmission is required and there are no core network delays. Moreover, LTE is also not prepared to operate at the required levels of reliability required in industrial environments.

5.3.2 WirelessHART and ISA100.11a

WirelessHART was developed to use the application layer of the Highway Addressable Remote Transducer (HART) protocol, which is an open protocol for communication in industrial environments using wired 4–20 mA lines Nixon and Round Rock. In addition, HART became widely spread worldwide due to its ease of deployment over existing cables in industrial installations.

WirelessHART uses an enhanced version of IEEE 802.15.4 for wireless communication and thus communicates at the unlicensed 2.4 GHz frequency band Nixon and Round Rock. It uses a mesh topology with redundancy and self-healing capabilities. Regarding the MAC layer, it uses time division multiple access (TDMA) with slow frequency hopping and allocates 10 ms slots for grant based communication. Moreover, it groups these slots into superframes, which are repeated cyclically. Lastly, to improve its performance, it also offers the possibility of blacklisting channels, in order to avoid interference.

Another technology similar to wirelessHART is ISA100.11a. It is an open standard developed by the International Society of Automation (ISA) and also relies on using unlicensed bands and operating with TDMA Nixon and Round Rock. However, despite their wide adoption, the fact that both wirelessHART and ISA100.11a use an unlicensed spectrum makes them unable to guarantee the target reliability of URLLC applications.

5.4 Enabling Technologies

In this section we present technologies that are being developed towards enabling URLLC capabilities in 5G networks. Each solution tackles a different source of delay and some of these solutions can be used in conjunction in order to deliver the target latency.

5.4.1 Faster Channel Coding

The time to decode the messages is a challenging aspect of URLLC, as by using simpler channel codes to increase the speed of decoding may result in a poorer performance in terms of reliability. This is a hot topic in URLLC research Shirvanimoghaddam et al., but more research is still needed.

An analysis of the complexity of existing channel coding algorithms has been presented in Sybis et al. (2016) and several candidates are compared. They conclude that the information block size greatly influences the performance of the coding scheme and that the optimal code to be used depends on it. Polar codes, which are far less complex to decode, perform well under short information block sizes, as do low-density parity-check (LDPC) codes, which are more complex. On the other hand, for medium-high length block sizes turbo codes are still the preferred choice in terms of performance Sybis et al. (2016).

Main research directions in terms of coding schemes towards URLLC include developing low complexity ordered statistics decoders, designing self-adaptive joint coding and modulation schemes and space-frequency channel coding Shirvanimoghaddam et al..

5.4.2 Latency Aware HARQ

In current technologies hybrid automatic repeat request (HARQ) is used to greatly improve the spectral efficiency of communication Rosas et al. (2016). HARQ is a retransmission mechanism that utilizes a feedback channel to let the transmitter know when the message has failed to be delivered. However, it may incur added latency to the application, which has to be overcome when considering URLLC applications.

On the other hand, HARQ constitutes an interesting method of using wireless resources, as it provides high diversity orders at low cost, since in most cases the average number of transmissions is very close to one Nadas et al. (2019). This is a strong motivation that has led researchers to investigate means of enabling its use even in URLLC applications.

There are several modifications that need to be made to traditional HARQ, as it is used in LTE, so that it can be used in URLLC scenarios. Firstly, the scheduling must be resolved, as currently the LTE scheduling mechanism is an intolerable source of delay. One solution is to operate on a grant-free basis, such that the receiver does not have to wait for a resource block to send the feedback messages and the transmitter also does not have to wait to send the retransmission. Network slicing solutions are ideal to fulfil this purpose.

Another source of delay that needs to be overcome in HARQ is the time to decode the messages. Traditional turbo codes used today in LTE are very robust in terms of bit error rate performance; however, messages sent using turbo encoding require a large number of operations to be decoded. By using faster codes as proposed above, this issue is minimized; however, there are other things that can be done on top of that, such as trying to predict the failure of a message before it is decoded. One family of solutions known as early HARQ (E-HARQ) rely on the detection or prediction of errors. For instance, it is possible to use the log likelihood ratios of the early stages of turbo encoding to get an estimate of whether or not the decoding will fail Berardinelli et al. (2016). Alternatively, LDPC codes (which have similar complexity to turbo codes Shirvanimoghaddam et al.) can be designed as an ensemble of sub-codes, which allow for early detection Goektepe et al. (2018). Moreover, machine learning methods can be coupled with the aforementioned strategies and provide

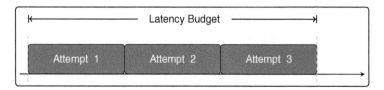

Figure 5.4 Latency aware HARQ.

an even better result in terms of early error detection Strodthoff et al. (2018).

Yet another issue with HARQ is that when retransmissions are allowed, the transmitter should convey the message in a fraction of the latency budget, to allow for the possibility of failure. One attractive approach is to increase the communication rate of each attempt Nadas et al. (2019) and therefore have all attempts fit within the latency budget. By setting the rate proportionally to the number of allowed attempts, it is possible to achieve the desired reliability and latency using HARQ. This idea is illustrated in Figure 5.4.

5.4.3 Joint Design

Joint design strategies are the ones where the application is designed alongside the communication link in order to obtain a better performance. When considering networked control systems (NCS), joint design strategies are one method for achieving URLLC Tong et al. (2018).

There are several means of communication and control joint design strategies that can be adopted. For instance, it is possible to optimize the energy efficiency of a sensor–controller link by setting different sampling frequencies alongside latency and reliability parameters depending on the control cost Park et al. (2011). In another scenario, such as complex industrial plants containing several control loops, it is possible to minimize the total control cost of the plants by jointly setting their sampling frequency and network parameters Lu et al. (2016).

As mentioned before, one particularly interesting class of controllers to be used with joint design techniques are MPCs, and more specifically, PPC. In the context of joint design, the following trade-off

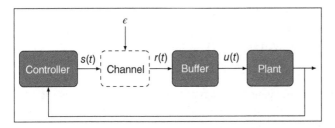

Figure 5.5 Control architecture with a wireless link connecting controller and actuator considering ϵ packet drop probability.

arises: reducing the length of the prediction horizon implies shorter messages (and thus easier to convey) but at the same time incurs fewer chances of missing the message. Thus, an optimal policy can be found with the objective of reducing wireless resource consumption or the control cost Tong et al. (2018).

Also regarding PPC, it can be combined with Chase combining HARQ (CC-HARQ) Chase (1985) to provide array gain on top of the diversity gain. In other words, failed messages would be retransmitted and combined at the receiver via maximum ratio combining (MRC), thus increasing the chances of detection.

Next we will present an analysis of how combining PPC with MRC can be used to reduce the system bandwidth. The topology considered here is that of Figure 5.2, wherein only the link between the controller and the plant is wireless and, as mentioned before, where spectral efficiency is an interesting metric to improve.

Consider a networked control architecture with a wireless link connecting the controller to the actuator – which is positioned at the plant – as illustrated in Figure 5.5. PPC is performed such that a buffer stores the vector $u(t)$ which contains K control signals to be used if packets are dropped.

In PPC, sending longer packets to include the additional control signals creates a trade-off, as it increases the message length (which imposes higher data rates for the same latency) resulting in a higher probability of communication error. At the same time, more control signals make it more fault tolerant. The stability and performance of the NCS depend on the packet drop rate ϵ_c and the communication latency λ.

5.4.3.1 Communication Model

At every instant t a packet containing K control signals is sent to the buffer. The received signal $r(t)$ is expressed by

$$r(t) = h(t)s(t) + w(t), \qquad (5.1)$$

where $s(t)$ is the generated signal at the controller, $w(t)$ is the additive white Gaussian noise (AWGN) and $h(t)$ is the channel gain, which is random and follows a Nakagami m distribution. We are assuming that the channel does not vary between symbols. Also, slow frequency hopping is used, such that subsequent transmissions are performed in uncorrelated frequency channels and thus the channel model is block fading.

The probability of dropping a packet ϵ depends on the condition of the wireless channel. As there is a strict deadline, the capacity at finite block length should be considered to obtain achievable rates. However, since the channel is block fading and rates are relatively high, as is the signal to noise ratio (SNR), average error probabilities are well approximated by outage probabilities Mary et al. (2016).

Considering L_H and L_U, the header and control signal lengths, the length of the forward packet is expressed via

$$L_T = L_H + KL_U. \qquad (5.2)$$

Lastly, the minimum bit rate R_b to meet the target latency is

$$R_b = \frac{L_T}{\lambda}. \qquad (5.3)$$

We use the minimum bit rate because using a faster one would result in a higher error probability for the same bandwidth B.

5.4.3.2 Proposed Solution

The proposed solution consists of the controller attempting to send a packet containing K control signals every λ s. If the packet fails, a retransmission is attempted using the CC-HARQ strategy and this is repeated up to z times. This is enabled by the fact that, in PPC, at any sampling instant t, K control signals are transmitted. Thus, if a message is decoded at instant $t + j$, the controller can obtain $K - j$ relevant control signals. The scheme is illustrated in Figure 5.6.

The maximum value of z that guarantees the availability of every control signals is $\lceil K/2 \rceil$. This is because using $z > \lceil K/2 \rceil$ violates the target error probability since when a message is decoded after having failed $j > \lceil K/2 \rceil$ times, only $K - j$ control signals are not outdated and the next packet would have to succeed in $K - j$ attempts, which is less than $\lceil K/2 \rceil$ by definition.

By carefully choosing the value for z and K, it is possible to optimize communication resources while still meeting the control design target. Since energy is not critical at the controller or the actuator, the focus of this work is to minimize the required bandwidth B for each transmission[1].

The outage probability $P_{\text{out},z}$ after z rounds in CC-HARQ is approximated at high SNR by Aalo (1995)

$$P_{\text{out},z} \approx \frac{\left(\frac{m\gamma_0}{\overline{\gamma}}\right)^{mz}}{\Gamma(mz + 1)}, \tag{5.4}$$

where m is the Nakagami m parameter, $\gamma_0 = 2^{Rb/B} - 1$ is the outage threshold from the channel capacity, $\Gamma(.)$ is the gamma function and the average SNR $\overline{\gamma}$ is obtained via

$$\overline{\gamma} = \frac{P_t}{BN_0 M_l A_0 d^\alpha}. \tag{5.5}$$

In (5.5), P_t is the transmit power, N_0 is the noise spectral power density, A_0 is the loss at a reference distance, d is the distance between transmitter and receiver, α is the path loss exponent and M_l is a link margin that accounts for any unforeseen losses and the noise figure. Since a minimum error probability must be guaranteed, $P_{\text{out},z} \leq \epsilon_c$ must hold. Thus, using (5.4) and (5.5) we arrive at

$$B\left(2^{\frac{R_b}{B}} - 1\right) \leq \beta, \tag{5.6}$$

where $\beta = \frac{(\epsilon_c \Gamma(mz+1))^{1/mz} P_t}{mN_0 M_l A_0 d^\alpha}$. Using W_{-1}, the lower part of the main Lambert W function, (5.6) becomes

$$B \geq \frac{-R_b \ln(2)}{W_{-1}(\frac{-R_b \ln(2)}{\beta})}. \tag{5.7}$$

Since W_{-1} can only yield real values for inputs greater than or equal to $-1/e$ (where e is Euler's constant), using (5.3) and analyzing the

---

1 The excess bandwidth consumed due to slow frequency hopping between different transmission attempts is not an issue since in CC-HARQ the average number of retransmissions is typically close to one, when the target error probability is very low, as in URLLC.

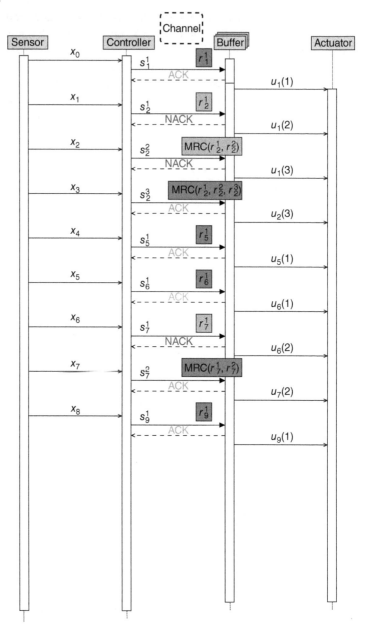

Figure 5.6 Proposed scheme, increasing the array gain by using CC-HARQ in conjunction with PPC.

argument of W_{-1} in (5.7) provides the bound

$$K \le \frac{1}{L_U} \left(\frac{\beta\lambda}{e \ln(2)} - L_H \right). \tag{5.8}$$

Finally, we write the optimization problem as

$$\underset{K \in \mathbb{N}^*, z \in \mathbb{N}^*}{\text{minimize}} \quad B = \frac{-R_b \ln(2)}{W_{-1}(\frac{-R_b \ln(2)}{\beta})} \tag{5.9a}$$

$$\text{subject to} \quad z \le [K/2], \tag{5.9b}$$

which can be tackled numerically.

5.4.3.3 Numerical Results and Conclusion

To illustrate the effect of K and z in the required bandwidth, we have performed numerical simulations computing B from (5.9a). The values used in the simulation are summarized in Table 5.1. Figure 5.7 shows the minimum required bandwidth for different values of K. Each curve represents a value of z, ranging from 2 to 5. Using the inequality in (5.8), we can observe that K would have to assume a negative value for $z = 1$, which is not possible. Here the best choice of parameters is $K = 5$ and $z = 3$, which shows how joint design can truly achieve better performance then considering only one aspect of the system.

Table 5.1 Simulation parameters.

Parameter	Value
Target error probability ϵ	10^{-9}
Maximum latency (λ)	1 ms
Transmit power (P_t)	-30 dB
Distance (d)	10 m
Header length (L_H)	40 bits
Control signal length (L_U)	16 bits
Link margin (M_l)	10 dB
Spectral noise power density (N_0)	-204 dB
Attenuation at 1 m (A_0)	30 dB
Path loss exponent (α)	3
Nakagami m parameter (m)	1

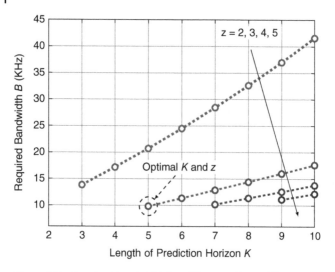

Figure 5.7 Required bandwidth for different values of *K* and *z*.

5.5 Conclusion

URLLC is an enabler of several core factory automation applications: however the existing network technologies have to be improved in order to support its stringent latency and reliability requirements. There are several interesting research directions such as communication and control joint design, latency aware HARQ, channel coding techniques and network slicing, which are bringing URLLC closer to reality.

Moreover, the flexible structure of 5G is going to allow for interesting business models to arise and will quickly provide the means for spreading and evolving technology through new business models.

Bibliography

3GPP. Study on latency reduction techniques for LTE. Technical Report (TR) 36.881, 3rd Generation Partnership Project (3GPP), 07 2016. URL http://www.3gpp.org/DynaReport/36881.htm. Version 14.0.0.

3GPP. Study on new radio access technology Physical layer aspects. Technical Report (TR) 38.802, 3rd Generation Partnership Project

(3GPP), 09 2017. URL http://www.3gpp.org/DynaReport/38802.htm. Version 14.2.0.

Valentine A Aalo. Performance of maximal-ratio diversity systems in a correlated Nakagami-fading environment. *IEEE Trans. Commun.*, 43 (8):2360–2369, 1995. doi: 10.1109/26.403769.

S. Ashraf, Y. P. E. Wang, S. Eldessoki, B. Holfeld, D. Parruca, M. Serror, and J. Gross. From radio design to system evaluations for ultra-reliable and low-latency communication. In *Proc. European Wireless 2017; 23th European Wireless Conf*, pages 1–8, May 2017.

G. Berardinelli, S. R. Khosravirad, K. I. Pedersen, F. Frederiksen, and P. Mogensen. On the benefits of early HARQ feedback with non-ideal prediction in 5G networks. In *Proc. Int. Symp. Wireless Communication Systems (ISWCS)*, pages 11–15, September 2016. doi: 10.1109/ISWCS.2016.7600846.

D. Chase. Code combining - a maximum-likelihood decoding approach for combining an arbitrary number of noisy packets. *IEEE Trans. Commun.*, 33(5):385–393, May 1985. ISSN 0090-6778. doi: 10.1109/TCOM.1985.1096314.

H. Chen, R. Abbas, P. Cheng, M. Shirvanimoghaddam, W. Hardjawana, W. Bao, Y. Li, and B. Vucetic. Ultra-reliable low latency cellular networks: Use cases, challenges and approaches. *IEEE Commun. Mag.*, 56(12):119–125, December 2018. ISSN 0163-6804. doi: 10.1109/MCOM.2018.1701178.

D. Cottet, W. v. d. Merwe, F. Agostini, G. Riedel, N. Oikonomou, A. Rueetschi, T. Geyer, T. Gradinger, R. Velthuis, B. Wunsch, D. Baumann, W. Gerig, F. Wildner, V. Sundaramoorthy, E. Bianda, F. Zurfluh, R. Bloch, D. Angelosante, D. Dzung, T. Wien, A. E. Vallestad, D. Orfanus, R. Indergaard, H. Vefling, A. Heggelund, and J. Bradshaw. Integration technologies for a fully modular and hot-swappable MV multi-level concept converter. In *Proc. PCIM Europe 2015; Int. Exhibition and Conf. for Power Electronics, Intelligent Motion Renewable Energy and Energy Management*, pages 1–8, May 2015.

Shuguang Cui, A. J. Goldsmith, and A. Bahai. Energy-constrained modulation optimization. *IEEE Trans. Wireless Commun.*, 4(5): 2349–2360, September 2005. ISSN 1536-1276. doi: 10.1109/TWC.2005.853882.

Salah-Eddine Elayoubi, Jean-Sébastien Bedo, Miltiades Filippou, Anastasius Gavras, Domenico Giustiniano, Paola Iovanna, Antonio

Manzalini, Olav Queseth, Theodoros Rokkas, Mike Surridge, et al. 5g innovations for new business opportunities. In *Mobile World Congress*. 5G Infrastructure association, 2017.

S. Feliciano, H. Sarmento, and J. de Oliverira. Field area network in a MV/lv substation: A technical and economical analysis. In *Proc. IEEE Int. Conf. Intelligent Energy and Power Systems (IEPS)*, pages 192–197, June 2014. doi: 10.1109/IEPS.2014.6874178.

B. Goektepe, S. Faehse, L. Thiele, T. Schierl, and C. Hellge. Subcode-based early HARQ for 5G. In *Proc. IEEE Int. Conf. Communications Workshops (ICC Workshops)*, pages 1–6, May 2018. doi: 10.1109/ICCW.2018.8403491.

Natalie Larson, Džiugas Baltrunas, Amund Kvalbein, Amogh Dhamdhere, kc claffy, and Ahmed Elmokashfi. Investigating excessive delays in mobile broadband networks. In *Proceedings of the 5th Workshop on All Things Cellular: Operations, Applications and Challenges*, AllThingsCellular '15, pages 51–56, New York, NY, USA, 2015. ACM. ISBN 978-1-4503-3538-6. doi: 10.1145/2785971.2785980.

Chenyang Lu, Abusayeed Saifullah, Bo Li, Mo Sha, Humberto Gonzalez, Dolvara Gunatilaka, Chengjie Wu, Lanshun Nie, and Yixin Chen. Real-time wireless sensor-actuator networks for industrial cyber-physical systems. *Proc. IEEE*, 104(5):1013–1024, 2016. doi: 10.1109/JPROC.2015.2497161.

M. Luvisotto, Z. Pang, and D. Dzung. Ultra high performance wireless control for critical applications: Challenges and directions. *IEEE Trans. Ind. Informat.*, 13(3):1448–1459, June 2017. ISSN 1551-3203. doi: 10.1109/TII.2016.2617459.

P. Mary, J. M. Gorce, A. Unsal, and H. V. Poor. Finite blocklength information theory: What is the practical impact on wireless communications? In *IEEE Globecom Workshops*, pages 1–6, December 2016. doi: 10.1109/GLOCOMW.2016.7848909.

J. P. B. Nadas, O. Onireti, R. D. Souza, H. Alves, G. Brante, and M. A. Imran. Performance analysis of hybrid ARQ for ultra-reliable low latency communications. *IEEE Sensors J.*, page 1, 2019. ISSN 1530-437X. doi: 10.1109/JSEN.2019.2891221.

Mark Nixon and TX Round Rock. A comparison of wirelesshart and isa100. 11a. *White Paper*.

D. Orfanus, R. Indergaard, G. Prytz, and T. Wien. Ethercat-based platform for distributed control in high-performance industrial applications. In *Proc. IEEE 18th Conf. Emerging Technologies Factory*

Automation (ETFA), pages 1–8, September 2013. doi: 10.1109/ETFA.2013.6647972.

Pangun Park, José Araújo, and Karl Henrik Johansson. Wireless networked control system co-design. In *Networking, Sensing and Control (ICNSC), 2011 IEEE International Conference on*, pages 486–491. IEEE, 2011. doi: 10.1109/ICNSC.2011.5874926.

P Popovski, V Braun, HP Mayer, P Fertl, Z Ren, D Gonzales-Serrano, E Ström, T Svensson, H Taoka, P Agyapong, et al. Scenarios requirements and KPIs for 5G mobile and wireless system. *ICT-317669-METIS/D1. 1, ICT-317669 METIS project*, 2013.

Fernando Rosas, Richard Demo Souza, Marcelo E Pellenz, Christian Oberli, Glauber Brante, Marian Verhelst, and Sofie Pollin. Optimizing the code rate of energy-constrained wireless communications with HARQ. *IEEE Trans. Wireless Commun.*, 15(1):191–205, 2016.

Mahyar Shirvanimoghaddam, Mohamad Sadegh Mohamadi, Rana Abbas, Aleksandar Minja, Balazs Matuz, Guojun Han, Zihuai Lin, Yonghui Li, Sarah Johnson, and Branka Vucetic. Short block-length codes for ultra-reliable low-latency communications. *IEEE Xplore*, 57.

Nils Strodthoff, Barış Göktepe, Thomas Schierl, Cornelius Hellge, and Wojciech Samek. Enhanced machine learning techniques for early HARQ feedback prediction in 5G. *arXiv preprint arXiv:1807.10495*, 2018.

M. Sybis, K. Wesolowski, K. Jayasinghe, V. Venkatasubramanian, and V. Vukadinovic. Channel coding for ultra-reliable low-latency communication in 5G systems. In *Proc. IEEE 84th Vehicular Technology Conf. (VTC-Fall)*, pages 1–5, September 2016. doi: 10.1109/VTCFall.2016.7880930.

Xin Tong, Guodong Zhao, Muhammad Ali Imran, Zhibo Pang, and Zhi Chen. Minimizing wireless resource consumption for packetized predictive control in real-time cyber physical systems. In *2018 IEEE International Conference on Communications Workshops (ICC Workshops)*. IEEE, 2018. doi: 10.1109/ICCW.2018.8403546.

Lei Wang, Junyan Qi, Huijuan Jia, and Bin Fang. The construction of soft servo networked motion control system based on ethercat. In *Proc. The 2nd Conf. Environmental Science and Information Application Technology*, volume 3, pages 356–358, July 2010. doi: 10.1109/ESIAT.2010.5568340.

6

Anomaly Detection and Self-healing in Industrial Wireless Networks

Ahmed Zoha, Qammer H. Abbasi, and Muhammad A. Imran

James Watt School of Engineering, University of Glasgow, UK

6.1 Introduction

The increased demands of high throughput, coverage and end user quality of service (QoS) requirements, driven by ever increasing mobile usage, incur additional challenges for network operators. Fuelled by mounting pressure to reduce capital and operational expenditures (CAPEX and OPEX) and improve efficiency in legacy networks, the self-organizing network (SON) paradigm aims to replace the classic manual configuration, post deployment optimization and maintenance in cellular networks with self-configuration, self-optimization, and self-healing functionalities. A detailed review of the state-of-the-art SON functions for legacy cellular networks can be found in (Aliu et al., 2013). The main task within the self-healing functional domain is autonomous cell outage detection and compensation. Current SON solutions generally assume that the spatio-temporal knowledge of a problem that requires SON based compensation is fully or at least partially available; for example, the location of coverage holes, handover ping-pong zones or congestion spots are assumed to be known by the SON engine. Traditionally, to assess and monitor mobile network performance, manual drive tests have to be conducted. However, this approach cannot deliver stringent resource efficiency and low latency, and cannot be used to construct dynamic models to predict system behaviour in a live operation fashion.

Wireless Automation as an Enabler for the Next Industrial Revolution, First Edition.
Edited by Muhammad A. Imran, Sajjad Hussain and Qammer H. Abbasi.
© 2020 John Wiley & Sons Ltd. Published 2020 by John Wiley & Sons Ltd.

This is particularly true for a *sleeping cell* (SC) scenario, which is a special case of cell outage that can remain undetected and uncompensated for hours or even days, since no alarm is triggered for an operation and maintenance (O&M) system as described in (Hämäläinen et al., 2012). An SC either causes deterioration of the service level or a total loss of radio service in its coverage area, due to a possible software (SW), firmware or hardware (HW) problem. An SC can only be detected by means of manual drive tests or via subscriber complaints. These solutions are not only time and resource consuming but also require expert knowledge to troubleshoot the problem. As future cellular networks have to rely more and more on higher cell densities, manual or semi-manual detection of an SC can become a huge challenge. Therefore, automatic cell detection has become a necessity so that timely compensation actions can be triggered to resolve any issues. Once the outage is detected, the operator can achieve self-resilience to network outages by employing an intelligent self-healing mechanism. The area under outage can be compensated by reusing the network's own resources through adjusting the antenna downtilt and variations in transmit power. A control mechanism is required to optimize the system parameters and lead the system closer to the normal state with minimum convergence time.

A self-healing block in SON consists of two modules, namely cell outage detection (COD) and cell outage compensation (COC). COD aims to autonomously detect outage cells, i.e. cells that are not operating properly due to possible failures, e.g. external failure such as power supply or network connectivity, or even misconfiguration (Hämäläinen et al., 2012; Liao et al., 2012; Mueller et al., 2008). On the other hand, COC refers to the automatic mitigation of the degradation effect of the outage by appropriately adjusting suitable radio parameters, such as the pilot power, antenna elevation and azimuth angles of the surrounding cells for coverage optimization (Hämäläinen et al., 2012).

The reported studies in the literature that address the problem of COD are either based on quantitative models (Barco et al., 2005), which require domain expert knowledge, or simply rely on performance deviation metrics (Cheung et al., 2004). Until recently, researchers have applied methods from the machine learning domain such as clustering algorithms (Ma et al., 2013) as well as Bayesian networks (Khanafer et al., 2008) to automate the detection of faulty cell behaviour. Coluccia *et al.* (Coluccia et al., 2013) analysed the variations in the traffic profiles for 3G cellular systems to detect real-world traffic anomalies. In particular, the problem of sleeping cell detection has been addressed by constructing and comparing a visibility graph of the network using *neighbour cell list (NCL)* reports (Mueller et al., 2008).

Compared to the aforementioned approaches, the COD solution proposed in this chapter differs in various aspects. Our proposed COD framework adopts a model-driven approach that makes use of a mobile terminal assisted data gathering solution based on minimize drive testing (MDT) functionality (Hämäläinen et al., 2012) as specified by 3GPP. MDT functionality allows evolved node Bs (eNBs) to request and configure user equipment (UE) to report back the key performance indicators (KPIs) from the serving and neighbouring cells along with their location information. To accurately capture the network dynamics, we first collect UE reported MDT measurements and further extract a minimalistic KPI representation by projecting them onto a low-dimensional embedding space. We then employ these embedded measurements together with density and domain based anomaly detection models, namely *local outlier factor based detector* (LOFD) and *one class support vector machine based detector* (OCSVMD). We compare and evaluate the performance of these learning algorithms to autonomously learn the 'normal' operational profile of the network, while taking into consideration the acute dynamics of the wireless environment due to channel conditions as well as load fluctuations. The learned profile leverages the intrinsic characteristics of embedded network measurements to intelligently diagnose a sleeping/outage cell situation. To the best of our knowledge, no prior study examines the use of OCSVMD and LOFD in conjunction with embedded MDT measurements for autonomous cell outage detection. This is in contrast to state-of-the-art techniques that analyse one or two KPIs to learn the decision threshold levels and subsequently apply them for detecting network anomalies. In addition, the COD framework further exploits the geolocation associated with each measurement to localize the position of the faulty cell, enabling the SON to autonomously trigger cell outage compensation actions.

Once the outage is properly detected, an automatic COC scheme is required for coverage optimization in order to continue serving the UEs in the outage area. Considering the acute dynamics of the always varying wireless environment in general, and the high variability in terms of load fluctuations, in dense wireless deployments, we propose a fuzzy logic based reinforcement learning (RL) algorithm, which allows learning online, through interactions with the surrounding environment, the best possible policy to compensate the outage. In the literature, the fuzzy logic algorithm has been studied (Razavi et al., 2010) to address the problem of self-recovery in the case of cell outage in LTE networks. Moreover, fractional power control based approaches in conjunction with reinforcement learning algorithm (Dirani and Altman, 2011) have also been studied to address the problem of the near far effect by controlling the required signal to interference and noise ratio (SINR), in order to reduce the call blocking rate. Motivated by this, in our paper we propose a fuzzy RL based compensation scheme in order to minimize the interference caused by the compensating sites.

The main contribution of this chapter can be summarized as follows: firstly, we propose a novel COD framework that exploits recently introduced MDT functionality in conjunction with state-of-the-art machine learning methods, to detect and localize cell outages in an autonomous fashion. Secondly, we demonstrate the applicability of a fuzzy reinforcement learning based method to achieve autonomous self-recovery in case of network outages. Finally, the proposed solution is validated with simulations that are set up in accordance with 3GPP LTE standards. The remainder of this paper is structured as follows: Section 6.2 present the system architecture for the proposed self-healing framework. Section 6.3 provides a detailed discussion of a COD framework, which also includes a brief description of LOFD and OCSVMD techniques that are used to profile, detect and localize anomalous network behaviour. In Section 6.4, the COC scheme is explained, whereas in Section 6.5 we provide details of our simulation setup and evaluation methodology. Furthermore, we present extensive simulation results to substantiate the performance of our proposed self-healing framework. Finally, Section 6.6 concludes this chapter.

6.2 System Design

To alleviate network performance deterioration, the first step is to detect the cell/base station (BS) in outage. This can be achieved by monitoring deviations from the KPI measurement report of the fault free network. Thereafter, the parameters of BSs neighbouring the outage BS, can be adjusted according to the operators policy so as to compensate for the outage situation. Hence, we propose a self-healing framework that primarily consists of the COD and COC stages, as illustrated in Figure 6.1.

6.2.1 COD Stage

As shown in Figure 6.1, firstly, to profile the normal operational behaviour of the network, our solution collects KPIs from the network leveraging MDT functionality. The goal is to use the learned profile to perform problem identification and localization autonomously during the monitoring period.

The MDT reporting schemes were defined in LTE Release 10 in 2011 (Hämäläinen et al., 2012). The release proposes constructing a database of MDT reports from the network using an *immediate* or *logged* MDT reporting configuration. In this study, the UEs are configured to report the cell identification and radio-measurement data to the target eNB based on the immediate MDT configuration procedure as shown in Figure 6.1. The signalling flow of MDT reporting procedure consists of configuration, measurement, reporting and storing phases. The UE is first configured to perform measurements periodically as well as whenever an A2 event (i.e. the serving cell becomes worst than the *threshold*) occurs. Subsequently, it performs KPI measurements: serving and neighbours' reference signal received power (RSRP), serving and neighbours' reference signal received quality (RSRQ), as specified in Table 6.1, and further reports it to the serving eNB. The eNB, after retrieving these measurements, further appends time and wideband channel quality information (CQI) and forwards it to a trace collection entity (TCE). The TCE collects and stores the trace reports, which are subsequently processed to construct a MDT database. In this study, the trace records obtained from the reference scenario (i.e. fault-free) act as benchmark data and are used by the anomaly detection models to learn the network

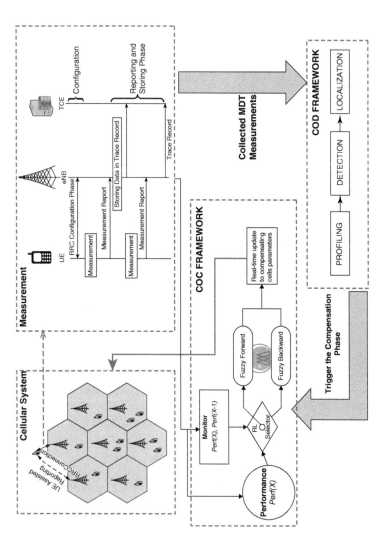

Figure 6.1 System model for autonomous COD and COC frameworks.

Table 6.1 MDT reported measurements.

Measurements	Description
Location	longitude and latitude information
Serving cell info	Cell global identity (CGI)
RSRP	Reference signal received power in dBm
RSRQ	Reference signal received quality in dB
Neighboring cell information	Three sStrongest intra-LTE RSRP, RSRQ information

profile. These models are then employed to autonomously detect and localize outage situations. The proposed framework as shown in Figure 6.1 consists of profiling, detection and localization phases, as detailed in Section 6.3.

6.2.2 COC Stage

Once the cell outage situation has been detected by the O&M, subsequently, a COC scheme is triggered to optimize the coverage and capacity of the identified outage zone according to the operator policies. This is achieved by adjusting the antenna gain through electrical tilt and the downlink transmission power of the potential compensating nodes. In our proposed framework, COC is implemented via a fuzzy logic based RL scheme, as illustrated in Fig 6.1, which is further explained in Section 6.4.

6.3 Cell Outage Detection Framework

The proposed COD framework consists of profiling, detection and localization phases, which are subsequently discussed in detail.

6.3.1 Profiling Phase

In the profiling phase, the trace records are processed to extract the feature vector **O** corresponding to each MDT measurement, as shown in Table 6.1. The measurements including reference signal received power and quality of the serving, as well as of the three strongest neighboring cells and the CQI are concatenated into a feature vector, **O**, which is expressed as follows:

$$\mathbf{O} = \left[RSRP_S, RSRP_{n1}, RSRP_{n2}, RSRP_{n3}, \right. \\ \left. RSRQ_S, RSRQ_{n1}, RSRQ_{n2}, RSRQ_{n3}, CQI \right] \tag{6.1}$$

where the subscripts S and n denote the serving and neighboring cells, respectively. The observation vector, **O**, is a nine-dimensional feature vector of numerical features that corresponds to one network measurement. To reduce the complexity of storage, processing and analysis this nine-dimensional feature vector is then embedded to three dimensions in the Euclidean space using the multi-dimensional scaling (MDS) method (Cox and Cox, 2010). MDS provides a low dimensional embedding of the target KPI vectors **O** while preserving the pairwise distances amongst them. Given, a $t \times t$ dissimilarity matrix $\mathbf{\Delta}^X$ of the MDT dataset, MDS attempts to find t data points $\psi_1, \ldots, \psi_t$ in m dimensions, such that $\mathbf{\Delta}^\Psi$ is similar to $\mathbf{\Delta}^X$. Classical MDS operates in Euclidean space and minimizes the following objective function

$$\min_{\psi} \sum_{i=1}^{t} \sum_{j=1}^{t} \left(\phi_{ij}^{(X)} - \phi_{ij}^{(\Psi)} \right)^2, \tag{6.2}$$

where $\phi_{ij}^{(X)} = \|x_i - x_j\|^2$ and $\phi_{ij}^{(\Psi)} = \|\psi_i - \psi_j\|^2$. Equation (6.2) can be reduced to a simplified form by representing $\mathbf{\Delta}^X$ in terms of a kernel matrix using

$$\mathbf{X}^T\mathbf{X} = -\frac{1}{2}\mathbf{H}\mathbf{\Delta}^X\mathbf{H}, \tag{6.3}$$

where $\mathbf{H} = \mathbf{I} - \frac{1}{t}\mathbf{e}\mathbf{e}^T$ and **e** is a column vector of all 1s. Hence (6.2) can be rewritten as

$$\min_{\psi} \sum_{i=1}^{t} \sum_{j=1}^{t} \left(x_i^T x_j - \psi_i^T \psi_j \right)^2. \tag{6.4}$$

As shown in (Cox and Cox, 2010), $\mathbf{\Psi}$ can be obtained by solving $\mathbf{\Psi} = \sqrt{\mathbf{\Lambda}}\mathbf{V}^T$, where **V** and $\mathbf{\Lambda}$ are the matrices of the top m eigenvectors and their corresponding eigenvalues of $\mathbf{X}^T\mathbf{X}$, respectively. The m-dimensional embedding of the data points are the rows of $\sqrt{\mathbf{\Lambda}}\mathbf{V}^T$, whereas the value of m is chosen to be 3 in our case. The preprocessing of the network observation $\mathbf{O}^e$ using the MDS method has several advantages. In the literature, the MDS technique has been widely used as a dimensionality reduction method (Cox and Cox, 2010) to transform high-dimensional data into a meaningful representation of reduced dimensionality. One of the problems with high-dimensional datasets is that in many cases not all of the measured variables are 'critical' for understanding the underlying phenomena. As shown in the literature, dimensionality reduction is a critical preprocessing step for the analysis of real-world datasets, since it mitigates the curse of dimensionality and other undesired properties of high-dimensional spaces. MDS aims to achieve an optimal spatial configuration in a low dimensional space such that distances in the new configuration (i.e. $\phi_{ij}^{(\Psi)}$) are close in value to the observed distances (i.e. $\phi_{ij}^{(X)}$). The spatial configurations helps to reveal a hidden structure that is not obvious from raw data matrices, allowing the exploration of the inter-relationships of high-dimensional spaces. Given the growing complexity of the networks, particularly in the case of SON, it is challenging to identify a few measurements that accurately capture the behaviour of the system. The MDS preprocessing of the network measurements allows achieving a reduced representation that corresponds to intrinsic dimensionality of the data. Consequently, the low-dimensional representation of network measurements facilitates data modelling and allows the anomaly detection algorithms to obtain a better estimation of data density. As a result, the anomalous network measurements can be detected with higher accuracy, as discussed below. Moreover, unlike other dimensionality reduction methods such as principal component analysis (PCA) or linear discriminant analysis, MDS does not make an assumption of linear relationships between the variables, and hence is applicable to a wide variety of data.

In addition to network measurements, each MDT report is tagged with the location and time information as listed in Table 6.1, which is used in conjunction with RSRP values to estimate the dominance or the coverage area of target BS in the network. The dominance map estimation is further used to autonomously localize the position of the outage BS.

The next step after the preprocessing is to develop a reference database, $\mathbf{D}_M$, by storing the embedded measurements that represent the normal operation of the network. As shown in Figure 6.2, this reference database is used by a state-of-the-art anomaly detection algorithm to learn the 'normal' network profile. The goal of these algorithms is to define an anomaly detection rule that can

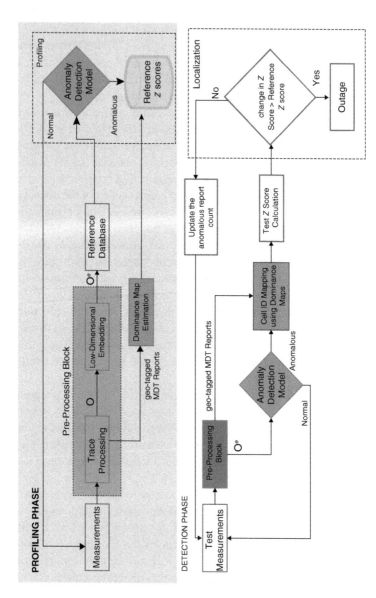

Figure 6.2 An overview of profiling and detection phases in the COD framework.

differentiate between normal and abnormal MDT measurements by computing a threshold 'φ' based on a dissimilarity measure '$\mathcal{D}$'. Thus, it can be treated as a binary classification problem that can formally be expressed as follows:

$$f(x_i) = \begin{cases} \text{Normal,} & \text{if } \mathcal{D}(x_i, \mathbf{D}_M) \leq \varphi \\ \text{Anomalous,} & \text{if } \mathcal{D}(x_i, \mathbf{D}_M) > \varphi \end{cases}. \tag{6.5}$$

Two state-of-the-art anomaly detection algorithms: OCSVMD and LOFD are examined for modelling the dynamics of network operational behaviour. A brief working description of the two detection algorithms are summarized as follows.

6.3.1.1 Local Outlier Factor Based Detector (LOFD)

The LOFD method (Breunig et al., 2000) adopts a density based approach to measure the degree of outlyingness of each instance. In comparison to nearest neighbour based approaches, it works by considering the difference in the local density ρ of the sample to that of its k neighbours, instead of relying on distance estimation alone. A higher score will be assigned to the sample if ρ is highly different from the local densities of its neighbour. The algorithm starts by first computing the distance of the measurement x to its kth nearest neighbour denoted by d_k, such that

$$d(x, x_j) \leq d(x, x_i) \quad \text{for at least } k \text{ samples}$$
$$d(x, x_j) < d(x, x_i) \quad \text{for at most } k - 1 \text{ samples.} \tag{6.6}$$

The subsequent step is to construct a neighbourhood $\mathcal{N}_k(x)$ by including all those points that fall within the d_k value. The following step is to calculate the reachability distance of sample x with respect to rest of the samples

$$d_r(x, x_i) = \max\{d_k(x_i), d(x, x_i)\}. \tag{6.7}$$

The local reachability density ρ is the inverse of average d_r and can be defined as

$$\rho(x) = \frac{|\mathcal{N}_k(x)|}{\sum_{x_i \in \mathcal{N}_k(x)} d_r(x, x_i)}. \tag{6.8}$$

Finally, the $S^{(\text{LOFD})}$ represents a local density-estimation score whereas value close to 1 mean x_i has same density relative to its neighbours. On the other hand, a significantly high $S^{(\text{LOFD})}$ score is an indication of anomaly. It can be computed as follows:

$$S^{(\text{LOFD})}(x) = \frac{\sum_{x_i \in \mathcal{N}_k(x)} \frac{\rho(x_i)}{\rho(x)}}{|\mathcal{N}_k(x)|}. \tag{6.9}$$

Since $S^{(\text{LOFD})}$ is sensitive to the choice of k, we iterate between $k_{\min}$ and $k_{\max}$ value for each sample, and take the maximum $S^{(\text{LOFD})}$ as described in Algorithm 1.

Algorithm 1 Local outlier factor based detection model.

1: Input Data $\mathcal{X} = \{x_j\}_{j=1}^N$, $k_{\min}$, $k_{\max}$
2: **for** $j = 1, 2, \ldots, N$: **do**
3:    **for** $k = k_{\min}$ to $k_{\max}$: **do**
4:       Find $d_k(x_j)$ from Equation 6.6
5:       Find the neighborhood $\mathcal{N}_k$ of x_j
6:       Calculate $d_r(x_j, x_i)$ from Equation 6.7
7:       Calculate $\rho(x_i)$ from Equation 6.8
8:       Calculate $S^{(\text{LOFD})}$ from Equation 6.9
9:    **end for**
10:   $S^{(\text{LOFD})} = \max(S^{(\text{LOFD})}_{k_{min}}, \ldots, S^{(\text{LOFD})}_{k_{max}})$
11: **end for**

6.3.1.2 One-Class Support Vector Machine based Detector (OCSVMD)

The one-class support vector machine by (Schölkopf et al., 2001) maps the input data/feature vectors into a higher dimensional space in order to find a maximum margin hyperplane that best separates the vectors from the origin. The idea is to find a binary function or a decision boundary that corresponds to a classification rule

$$f(x) = \langle \mathbf{w}, \mathbf{x} \rangle + b. \tag{6.10}$$

The $\mathbf{w}$ is a normal vector perpendicular to the hyperplane and $\frac{b}{\|\mathbf{w}\|}$ is an offset from the origin. For linearly separable cases, the maximization of margin between two parallel hyperplanes can be achieved by optimally selecting the values of w and b. This margin, according to the definition is $\frac{2}{\|\mathbf{w}\|}$. Hence, the optimal hyperplane should satisfy the following conditions minimize $\frac{1}{2}\|\mathbf{w}\|^2$

subject to : $y_i(\langle \mathbf{w}, x_i \rangle + b) \geq 1$ (6.11)

for $i = 1, \dots, N$.

The solution of the optimization problem can be written in an unconstrained dual form that reveals that the final solution can be obtained in terms of training vectors that lie close to the hyperplanes, also referred to as support vectors. To avoid overfitting on the training data, the concept of *soft decision* boundaries was proposed, and slack variable ξ_i and regularization constant v are introduced in the objective function. The slack variable is used to soften the decision boundaries, while v controls the degree of penalization of ξ_i. Few training errors are permitted if v is increased while degrading the generalization capability of the classifier. A *hard margin* SVM classifier is obtained by setting the value of $v = \infty$ and $\xi = 0$. The detail mathematical formulation for SVM models can be found in (Schölkopf et al., 2001). The original formulation of SVM is for linear classification problems; however non-linear cases can be solved by applying a kernel trick. This involves replacing every inner product of $x.y$ by a nonlinear kernel function, allowing the formation of nonlinear decision boundaries. The possible choices of kernel functions includes polynomial, Gaussian radial basis function (RBF) and sigmoid. In this study, we have used the RBF kernel: $\kappa(x, y) = \exp(-\|x - y\|^2 / 2\sigma^2)$, and the corresponding parameter values of the model are selected using cross validation method, as described in Algorithm 2.

As shown in Figure 6.2, using the benchmark data, we compute a reference z score for each target eNB in the network. The z score is calculated as follows: $z_b = \frac{|n_b - \mu_n|}{\sigma_n}$ where n_b is the number of MDT reports labelled as anomalies for the eNB b, and variables μ_n and σ_n are the mean and standard deviation anomaly scores of the neighbouring cells. In the profiling phase, we also estimate the so-called dominance area, i.e. for each cell, we define the area where its signal is the strongest. This is to establish the coverage range for each cell by exploiting the location information tagged with each UE measurement. The dominance estimation is required to determine a correct cell and MDT measurement association during an outage situation. This is because as soon as the SC situation triggers in the network, the malfunctioning eNB either becomes completely unavailable or experience severe performance issues. This triggers frequent UE handovers to the neighboring cells, and as a result the reported measurements from the affected area contains the neighbour cell E-UTRAN CGI, instead of the target cell. Hence, CGI alone cannot be used to localize the correct position of a faulty cell during an outage situation. The detection and localization phase of our COD framework make use of an estimated dominance map and reference z score information established in the profiling phase to detect and localize the faulty cell, as discussed in the following subsection.

6.3.2 Detection and Localization Phase

In the detection phase, the trained detection model is employed to classify network measurements as normal or anomalous. The output of the detection models allow us to compute a test z score for each eNB. To establish a correct cell measurement association, the geolocation of each report is correlated with the estimated dominance maps. In this way, we can achieve detection and localization by comparing the deviation of test z score of each cell with that of reference z score, as illustrated in Figure 6.2.

6.4 Cell Outage Compensation

The output of the detection phase is fed into the cell outage compensation module. This module is based on the combination of fuzzy logic and the RL algorithm (Saeed et al., 2012). In contrast to binary (0,1) decisions, fuzzy logic is a form of many-value logic and it represents uncertainty by outputting numeric values between 0 and 1. Such binary machine-like decision making is not always appropriate for applications where a more dynamic and human-like approach is needed. Fuzzy logic induces various degrees of outputs depending on the input conditions. The main benefits of such outputs are the controlling of a system that can be performed by using linguistic terms such as 'high' or 'low' instead of providing actual numerical values. Three main components of a fuzzy logic based system are fuzzification, rule-based inference and defuzzification. Fuzzification is the mapping of crisp input variables to fuzzy sets (linguistic interpretation), rule-based inference is the process of taking decisions based on 'if...then' pre-set rules and finally defuzzification process generates quantifiable crisp outputs based on the degree of membership of the outcome in the fuzzy sets. For our specific problem of multiple inputs and multiple outputs with a requirement of degree of membership function, Mamdani type fuzzy logic is preferred due to its simplicity and more suitable for multiple outputs, whereas Sugeno type fuzzy has a single output without membership functions. Further, we employ the *temporal difference* scheme to solve our problem, as it does not require modelling of the environment dynamics and can be implemented in an incremental fashion (Razavi and Claussen, 2013).

As for RL, it is a dynamic machine learning mechanism that interacts with the real time changes in the environment. The RL algorithm learns from its past experiences (exploitation) and new actions (exploration), unlike supervised learning where the system is explicitly taught. A new action is considered as a *reward*, if it generates a positive result towards the desired objective else the action is considered a penalty.

In our study, we demonstrate the combination of fuzzy rules in conjunction with the RL algorithm for COC. As shown in Figure 6.1, the objective is to improve the $\mathrm{Perf}(X)$ measurements. To achieve this, the proposed solution compares the current performance $\mathrm{Perf}(X)$ of the outage cell and its neighbours against their previous performance $\mathrm{Perf}(X-1)$. This change in performance is closely monitored by the reinforcement learning module to select the direction of the fuzzy module based on previous actions. Based on preset fuzzy module rules, the antenna downtilt and transmit powers of the neighbouring potential compensator are changed. After each cycle of action (change in antenna downtilt and transmit powers) the new accumulated $\mathrm{Perf}(X)$ is compared against the previous $\mathrm{Perf}(X-1)$ by the reinforcement learning algorithm. As shown in Figure 6.1, if the change is accessed as a *reward*, the fuzzy forward module is activated. Likewise, if the change is accessed as a *penalty*, the fuzzy backward module is activated for the next iteration. The fuzzy logic system iteratively reduces the resolution of the change in *action* as the target performance reaches closer to the performance when no outage was detected. This iterative process halts if their is no improvement detected in $\mathrm{Perf}(X)$ performance as compared to $\mathrm{Perf}(X-1)$ in either of the fuzzy directions. At this stage, we consider that the algorithm has finalised the best possible compensation parameters for the potential neighbouring cells.

6.5 Simulation Results

In this section, we demonstrate the performance of our COD framework by presenting the simulation results obtained under different network operating conditions.

6.5.1 Simulation Setup

To simulate the LTE network based on 3GPP specifications, we employ a fully dynamic system tool. We set up a baseline reference scenario that consists of 27 eNBs having an inter-site distance of 1000 m, with a cell load of 10 users. To model the variations in signal strength due to topographic features in an urban environment, the shadowing is configured to be 8 dB. Normal periodical MDT measurements as well as RLF-triggered data due to intra-network mobility, reported by UEs to eNBs, is used to construct a reference database for training outage detection models. To simulate a hardware failure in the network, at some point in the simulation the antenna gain of a BS is attenuated to −50 dBi, which leads to a cell outage in a network.

The measurements collected from the outage scenario are then used to evaluate the detection and localization performance of the proposed COD framework. In order to evaluate the performance of the compensation module, we identify three neighbouring sectors to compensate for the outage area. The antenna downtilt and transmission power of the neighbours is adjusted and optimised so that the best possible configuration is set to safeguard UEs in outage. The detailed simulation parameters are listed in Table 6.2. The detection performance of the outage detection models is also examined for different network configurations, obtained by varying the simulation parameter settings for ISD, load and shadowing.

6.5.1.1 Parameter Estimation and Evaluation

The parameter selection for LOFD and OCSVMD is performed using a combination of grid search and cross-validation (CV) method

Table 6.2 Simulation parameters.

Parameter	Values
Cellular layout	27 Macrocell sites
Inter-site distance (ISD)	1000m
Sectors	3 Sectors per cell
User distribution	Uniform random distribution
Path loss	L (dB)= $128.1 + 37.6\log_{10}(R)$
Antenna gain (normal scenario)	18 dBi
Antenna gain (SC scenario)	−50 dBi
Slow fading std	8 dB
Simulation length	420s (1 time step = 1 ms/1 TTI)
Control BS Tx power	46 dBm
Data BS Tx power	23 dBm
Horizontal HPBW	70°
Verticall HPBW	6.8°
Antenna pattern Saeed et al. (2012)	$B_\phi(\phi) = -\max(B_\phi, 12. \frac{\phi-\Phi}{\Delta_\phi})$
Network synchronization	Asynchronous
HARQ	Asynchronous, 8 SAW channels, max retransmission = 3
Cell selection criteria	Strongest RSRP defines the target cell
Load	20 users/cell
MDT reporting interval	240 ms
Traffic model	Infinite buffer
HO margin	3 dB

as listed in Algorithm 2. Initially, a grid of parameter values are specified that defines the parameter search space. For example, the hyper-parameters of OCSVM ν and kernel parameter γ is varied from 0 to 1 with 0.05 intervals to determine different combinations. Subsequently, for every unique parameter combination C_i, CV is performed as follows: the D_M is divided into training D_{train} and validation dataset D_{val}, and subsequently performance of the model is evaluated using the K folds approach as shown in Algorithm 2. The value of K

is chosen to be 10 in our framework. The performance estimate of the model over K folds is averaged and iteratively this process is repeated until all the parameter combinations are exhausted. C_i yielding the highest performance estimate is selected as an optimal parameter combination for the target model. The values of k_{min} and k_{max} for LOFD are found to be 5 and 14, respectively. In the case of OCSVMD, the RBF kernel is employed and the values of the hyperparameters v and γ are found to be 0.3 and 0.25, respectively. Finally, the test data D_{test} from the outage scenario has been used to estimate the performance of the trained models.

In our study, the quality of the target models is evaluated using receiver operating characteristic (ROC) curve analysis. The ROC curve plots the true positive rate or detection rate (DR) (i.e. a percentage of anomalous measurements correctly classified as anomalies) against the false positive rates (FPRs) (i.e. a percentage of normal cell measurements classified as anomalies) at various threshold settings. An area under ROC curve (AUC) metric is used for model comparison, whereas an AUC value of 1 or close to it, is an indicator of higher discriminatory power of the target algorithm.

Algorithm 2 Parameter estimation using the CV method.

1: Define parameter combination $C_i, i = 1, \dots, M$
2: **for** $i = 1, 2, \dots, M$: **do**
3:     Split the target dataset D_M into K chunks.
4:     **for** $l = 1, 2, \dots, K$: **do**
5:        Set D_{val} to be the lth chunk of data
6:        Set D_{train} to be the other $K - 1$ chunks.
7:        Train model using C_i, D_{train} and evaluate its performance P_l on D_{val}.
8:     **end for**
9:     Compute average performance P_i over K chunks
10: **end for**
11: **Parameter selection**: select C_i corresponding to highest P_i
12: **Performance estimation**: evaluate the performance of the model $M(C_i, D_M)$ on D_{test}

6.5.2   Cell Outage Detection Results

The training database D_M contains preprocessed embedded measurements from the reference scenario as discussed in Section 6.3.1. The database is subsequently used to model the normal operational behaviour of the network. The database measurement also includes RLF-triggered samples, since even in the reference scenario UEs experience connection failures due to intra-LTE mobility or shadowing. The test data collected from the outage scenario is used to evaluate the performance of the outage detection models.

The diagnosis process has been tested in twelve scenarios by changing the shadowing, user density and inter-site distance (ISD) parameters of the baseline simulation setup as listed in Table 6.2. We have evaluated the detection performance of the OCSVMD and LOFD against every target network configuration. Figure 6.3(a), illustrates the MDS projection of MDT measurements from the normal and the outage scenario using the baseline network operational settings. It can be observed that the abnormal measurements belonging to SC scenario lies far from the regular training observations. As discussed earlier in Section 6.3.1, MDS tries to maximize the variance between the data points and consequently dissimilar points are projected far from each other, allowing the models to compute a robust dissimilarity measure for outage detection. The goal of OCSVMD is to learn a close frontier delimiting the contour of training observations obtained from the non-outage scenario. In this way, any observation that lies outside of this frontier-delimited subspace (i.e. representative of the normal state of the network) is classified as an anomaly or an abnormal measurement. However, the inlier population (i.e. measurements that lie inside the OCSVMD frontier) is contaminated with RLF events, which ultimately elongates the shape of the learned frontier. As a result, during the detection phase, the observations from the outage scenario exhibiting similarity to RLF-like observations are positioned within the frontier-delimited space as shown in Figure 6.3(a), and hence wrongly classified as normal. The shape of the learned frontier determines the precision of the model for detecting anomalous network measurements.

To study the impact of different radio propagation environment on the detection performance, we varied the shadowing parameter

(a)

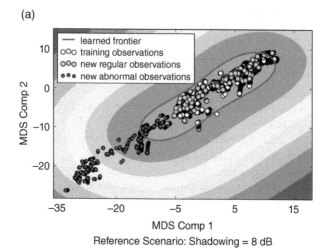

Reference Scenario: Shadowing = 8 dB

(b)

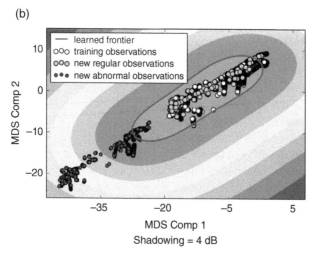

Shadowing = 4 dB

Figure 6.3 (a) OCSVMD learned network profile for reference scenario, (b) low-shadowing case, (c) distribution of RSRP values for all shadowing cases, (d) medium traffic case, (e) smaller ISD case, (f) distribution of RSRP values for all ISD cases.

from 8 dB to 4 dB and 12 dB cases. Under low-shadowing conditions (i.e. 4dB), it can be observed from Figure 6.3(b) that inlier population exhibits wider separation from anomalous observation in comparison to reference scenario. This is because higher shadowing

(c)

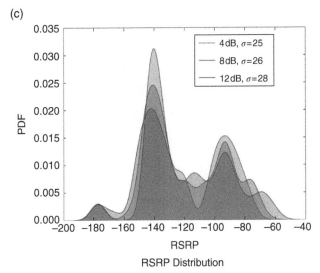

RSRP Distribution

(d)

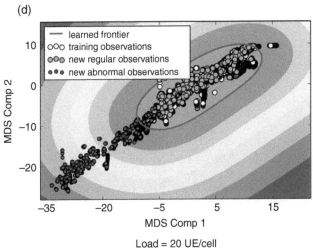

Load = 20 UE/cell

Figure 6.3 (Continued)

conditions affect the spread of the KPI measurements, as indicated in Figure 6.3(c). It can be inferred from the ROC analysis of OCSVMD, that detection performance deteriorates as the shadowing effect is varied from low to high. As shown in Figure 6.4(a), at target false positive rate of 10%, the model reports the highest detection rate

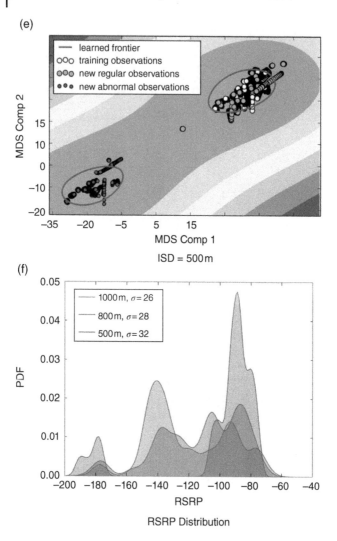

Figure 6.3 (Continued)

(i.e. TPR) of 93% under low-shadowing conditions. Likewise, the AUC score has also decreased from 0.98 to 0.94 for high-shadowing scenario (i.e. 12 dB). Moreover, we also analysed the OCSVMD detection performance under varying traffic conditions. Figure 6.3(d) depicts the distribution of measurements in the MDS space for a user

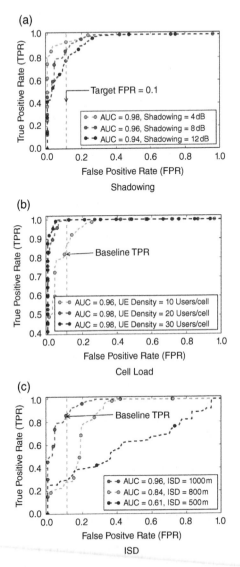

Figure 6.4 OCSVMD ROC curves for shadowing, traffic and ISD cases.

density of 20 per cell. The higher user density implies an increase in the number of training observations, which leads to a more accurate estimate of the frontier shape. This explains the slight improvement in the AUC score for OCSVMD with the increase in the cell load as shown in Figure 6.4(b). A notable detection rate improvement of 10% is observed for high traffic scenario (i.e. 30 users per cell) in comparison to the baseline OCSVMD.

As for different ISD configurations of a network, we see a significant change in the values of KPI measurements. This is expected since there is a strong correlation between UE reported KPIs and their distance from the eNB. Figure 6.3(f) shows the distribution of UE reported RSRP values for three different ISD cases. In the case of ISD= 500 m, we see a distinct peak of RSRP values around −90 dBm. Likewise, at the far left end we see a small peak around −180 dBm that is mainly due to RLF-like observations. In contrast, when ISD= 1000 m, the highest peak value is observed at around −140 dBm, and the observed measurements have lower data spread, as indicated in Figure 6.3(f). As already highlighted, the shape of the learned frontier by OCSVMD is directly affected by the distribution of observations in the embedded space. This becomes evident in Figure 6.3(e) which shows that the OCSVMD learns two decision frontiers instead of one, since there exist two distinct modes in the data distribution, for the case of ISD= 500 m. As a result, OCSVMD interprets a region where RLF-like events are clustered, as inliers, which leads to an inaccurate network profile. The ROC analysis shown in Figure 6.4(c) clearly indicates the degradation of OCSVMD performance for lower ISD values.

Similar to OCSVMD, the performance of LOFD is also evaluated for all target network configurations. As explained in Section 6.3.1, LOFD derives a measure of outlyingness of an observation (i.e. S^{LOFD}), based on the relative data density of its neighborhood. Figure 6.5(a) illustrates the labels assigned by LOFD to the observations obtained from the baseline scenario. It can be observed that LOFD even classifies some of the test instances that lie close to the vicinity of training observations as anomalous. Due to such instances LOFD receives a high outlying scores S^{LOFD}, since the local density around it is highly different from the density of its neighborhood. To further illustrate the impact of the variation and spread of the data on the values of S^{LOFD}, we plot a cumulative distribution function (CDF) for different shadowing scenarios, as shown in Figure 6.5(b). It can be seen that for the low-shadowing scenario almost 80% of the observations obtain a S^{LOFD} value less than 50. However, as the shadowing increases we see

Figure 6.5 Network profiling using LOFD.

(a)

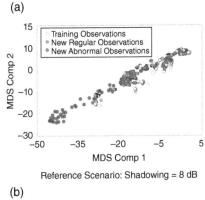

Reference Scenario: Shadowing = 8 dB

(b)

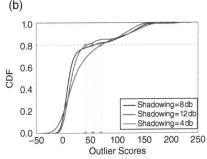

CDF of S^{LOFD} for Shadowing cases

(c)

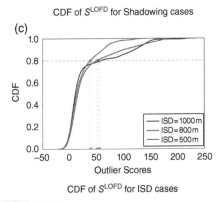

CDF of S^{LOFD} for ISD cases

a gradual increase in the value of S^{LOFD}. Likewise, a similar behaviour is observed with the increase in ISD, as shown in Figure 6.5(c). The shadowing and ISD parameters influence the distribution and spread of the data, as explained earlier, and consequently the value of S^{LOFD}.

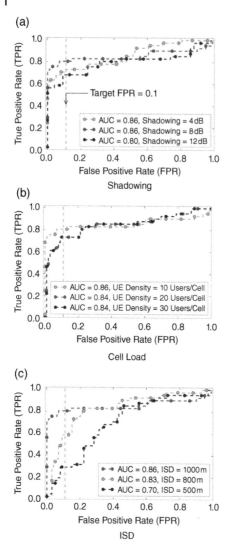

Figure 6.6 LOFD ROC curves for shadowing, traffic and ISD cases.

This leads to a low detection performance of LOFD, since it generates an increased number of false alarms.

As shown in Figure 6.6(a), the AUC score for LOFD decreases for the high-shadowing scenario. On the other hand, an increase in the cell load also increases the spread of the data, which consequently affects the detection performance of LOFD. As shown in Figure 6.6(b), at a false alarm rate of 10%, the highest detection rate of 81% is achieved for a network scenario in which load configuration is set to be 10 users per cell. Similarly, the change in the ISD has a severe effect on the model performance and low detection performance of 60% and 30% is achieved for 800 m and 500 m ISD configurations, as shown in Figure 6.6(c).

In summary, we can conclude from the reported results that OCSVMD under most cases achieves a better detection performance in comparison to LOFD. The outage detection models yield worst performance scores, particularly for low ISD network configuration. The performance issue of the target outage detection models can be addressed as follows: for OCSVMD, in the preprocessing step the RLF-like events must be filtered before constructing a training database. This would help decrease the spread of the data and the model would only learn the frontier that corresponds to normal operational network behaviour. In the case of LOFD, incremental drift detection schemes can be incorporated to re-tune the model parameters in order to minimize the false alarm rate.

6.5.3   Localization

Since the OCSVMD model has outperformed LOFD for most test cases, it has been selected as a final model to compute per cell z scores for the normal and SC scenarios, as shown in Figure 6.7. It can be observed from Figure 6.7 that measurements are classified as anomalous even in the normal operational phase of the network due to the occurrence of RLF events. This is particularly true for cell IDs 1, 5, 11, 16 and 19 whose n_b values are found to be 700, 2000, 3000, 1500, and 1200, respectively in the reference scenario. However, during an outage scenario, since cell 11 is configured as a faulty cell, the corresponding z scores are significantly higher than the rest of the network. A simple decision threshold can be applied on the computed z scores to autonomously localize faulty cells, and consequently, an alarm can be triggered. In addition to cell outage localization, the change in the z score can be used to identify performance degradation issues or a weaker coverage problems. This information can act as an input to the self-healing block of SON engine, which can then trigger the automated recovery process.

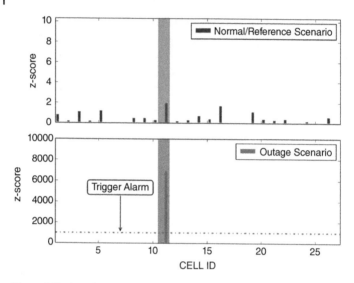

Figure 6.7 Localization of SC based on per cell *z* scores.

6.5.4 Compensation

The plots shown in Figures 6.8(a), 6.8(b) and 6.8(c) present the radio environment maps (REMs) for the normal, outage and compensated cases respectively, and the colour bar shows the SINR levels. As the subject cell site goes into outage, it is visible that the coverage area of the outage cell has very low SINR levels (depicted in 6.8(b)). The UEs in this low SINR region are susceptible to outage (failure of link to the network). The compensator module optimizes the antenna downtilt and transmit powers of the three potential neighbors such that this coverage gap is filled. It is visible from REM in Figure 6.8(c) that SINR of outage region is significantly improved after compensation.

Figure 6.8(d) presents the comparison of SINR levels with a CDF plot for the region in and around the outage cell. It is visible in the zoomed figure that, in the outage case, there are several UEs in the low SINR region. Whereas after compensation there is a significant reduction in the percentile of users in the low SINR region. Another visible effect is that in the compensation case a majority of the UEs also have substantially high SINR levels. This is due to the fact that the increase in the transmit power and change in antenna downtilt configuration further improves the SINR performance of the UEs closer to the compensating neighbors.

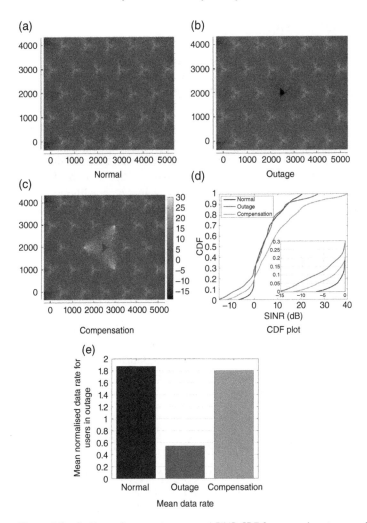

Figure 6.8 Radio environment maps and SINR CDF for normal, outage and compensated cases.

We also present in Figure 6.8(e) the bandwidth normalised data rate performance of only the UEs present in the outage area. It is visible that the mean data rate performance of the UEs is significantly reduced in the case of an outage. However, as the compensation is applied, this performance is improved upto a level negligibly less than the normal condition.

6.6 Conclusion

This chapter has presented a data-driven analytics framework for autonomous outage detection and coverage optimization in an LTE network that exploits the minimization of drive test functionality as specified by 3GPP in Release 10. The outage detection approach first learns a normal profile of the network behaviour by projecting the network measurements to a low-dimensional space. For this purpose, the multi-dimensional scaling method in conjunction with domain and density based detection models, OCSVMD and LOFD, respectively, are examined for different network conditions. It is established that OCSVMD, a domain based detection model attained a higher detection accuracy compared to LOFD, which adopts a density based approach to identify abnormal measurements. Finally the UE reported coordinate information is employed to establish the dominance areas of target cells that are subsequently used to localize the position of the outage zone. To optimize the coverage and capacity of the identified outage zone, a fuzzy based RL algorithm for COC is proposed. The COC algorithm achieves coverage optimization by adjusting the gains of the antennas through electrical tilt, and downlink transmission power of the neighbouring BSs. Simulation results have shown that the COC algorithm can recover a significant number of UEs from outage.

Bibliography

Osianoh Glenn Aliu, Ali Imran, Muhammad Ali Imran, and Barry Evans. A survey of self organisation in future cellular networks. *IEEE Communications Surveys & Tutorials*, 15(1):336–361, 2013.

Raquel Barco, Volker Wille, and Luis Díez. System for automated diagnosis in cellular networks based on performance indicators. *European Transactions on Telecommunications*, 16(5):399–409, 2005.

Markus M Breunig, Hans-Peter Kriegel, Raymond T Ng, and Jörg Sander. Lof: identifying density-based local outliers. In *ACM Sigmod Record*, volume 29, pages 93–104. ACM, 2000.

Benjamin Cheung, Stacy Gail Fishkin, Gopal N Kumar, and Sudarshan A Rao. Method of monitoring wireless network performance, September 21 2004. US Patent App. 10/946,255.

A Coluccia, A D'Alconzo, and F Ricciato. Distribution-based anomaly detection via generalized likelihood ratio test: A general maximum entropy approach. *Computer Networks*, 57(17):3446–3462, 2013.

Trevor F Cox and Michael AA Cox. *Multidimensional scaling*. CRC Press, 2010.

Mariana Dirani and Zwi Altman. Self-organizing networks in next generation radio access networks: Application to fractional power control. *Computer Networks*, 55(2):431–438, 2011.

Seppo Hämäläinen, Henning Sanneck, Cinzia Sartori, et al. *LTE Self-Organising Networks (SON): Network Management Automation for Operational Efficiency*. John Wiley & Sons, 2012.

Rana M Khanafer, Beatriz Solana, Jordi Triola, Raquel Barco, Lars Moltsen, Zwi Altman, and Pedro Lazaro. Automated diagnosis for umts networks using bayesian network approach. *IEEE Transactions on Vehicular Technology*, 57(4):2451–2461, 2008.

Qi Liao, Marcin Wiczanowski, and Slawomir Stanczak. Toward Cell Outage Detection with Composite Hypothesis Testing. In *IEEE ICC*, pages 4883–4887, June 2012.

Yu Ma, Mugen Peng, Wenqian Xue, and Xiaodong Ji. A dynamic affinity propagation clustering algorithm for cell outage detection in self-healing networks. In *Proceedings of IEEE Wireless Communications and Networking Conference (WCNC)*, pages 2266–2270. IEEE, 2013.

Christian M Mueller, Matthias Kaschub, Christian Blankenhorn, and Stephan Wanke. A cell outage detection algorithm using neighbor cell list reports. In *Self-Organizing Systems*, pages 218–229. Springer, 2008.

R. Razavi and H. Claussen. Improved fuzzy reinforcement learning for self-optimisation of heterogeneous wireless networks. In *Telecommunications (ICT), 2013 20th International Conference on*, pages 1–5, May 2013.

Rouzbeh Razavi, Siegfried Klein, and Holger Claussen. A fuzzy reinforcement learning approach for self-optimization of coverage in lte networks. *Bell Labs Technical Journal*, 15(3):153–175, 2010.

A. Saeed, O.G. Aliu, and M.A. Imran. Controlling self healing cellular networks using fuzzy logic. In *Wireless Communications and Networking Conference (WCNC), 2012 IEEE*, April 2012.

Bernhard Schölkopf, John C Platt, John Shawe-Taylor, Alex J Smola, and Robert C Williamson. Estimating the support of a high-dimensional distribution. *Neural computation*, 13(7):1443–1471, 2001.

7

Cost Efficiency Optimization for Industrial Automation

Hafiz Husnain Raza Sherazi[1], Luigi Alfredo Grieco[1], Gennaro Boggia[1], and Muhammad A. Imran[2]

[1] Department of Electrical and Information Engineering, Politecnico di Bari, Italy
[2] James Watt School of Engineering, University of Glasgow, UK

7.1 Introduction

The recent advancements in internet technologies have led automation processes towards the fourth industrial revolution. Cyber physical systems are the core of industry 4.0 and consistently monitor the physical processes in an industry, collecting real time data and implementing a distributed decision-making approach based on that data. This way, industry is capable of avoiding several types of unwanted costs by employing predictive maintenance (PM), anomaly detection, and condition monitoring systems.

The Internet of Things (IoT) (Atzori et al., 2010b; Gubbi et al., 2013; Weber and Weber, 2010) can generally be seen as a networked system of smart interconnecting objects (i.e. sensors, actuators, machines, vehicles, smart phones, and tablets, etc.)(Hersent et al., 2012). These devices are capable of sensing different types of data (i.e. temperature, pressure, light, acceleration, and so on) from the environment and, simultaneously processing it in real time in order to react to some particular events.

The anticipated benefits gained by the rise of the IoT are significant in terms of economic impact and improvement of quality of life (Grieco et al., 2014). As per some recent statistics forecast by some industrial giants (Emmerson, 2017; Ericsson, 2018; Evans, 2011)

Wireless Automation as an Enabler for the Next Industrial Revolution, First Edition.
Edited by Muhammad A. Imran, Sajjad Hussain and Qammer H. Abbasi.

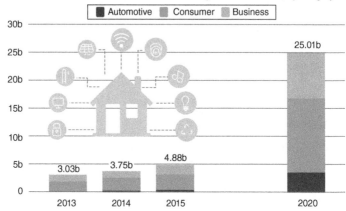

Figure 7.1 Evolution of the number of IoT devices and world population in the last decade.

around 25 billion connecting devices are expected by 2020 as a part of the IoT, more than six times the world's population, as depicted in Figure 7.1. According to an estimate, the total number of IoT devices in use reached 8.4 billion in 2017, a 31% increase as compared to 2016.

On the other hand, Intel projected the penetration of internet enabled devices to grow from 2 billion in 2006 to 200 billion by 2020[1], which implies nearly 26 smart devices for each human on Earth. As per another estimation[1], the number of connected devices will be 75.4 billion in 2025 and 125 billion by 2030. Some companies have presented their numbers taking smartphones, tablets, and computers out of the equation. Gartner estimated, 20.8 billion connected things will be in use by 2020. Similarly, IDC estimated this number to be 28.1 billion and BI Intelligence estimated it to be 24 billion. Gartner also gave an estimate of the total amount spent on IoT devices and services at nearly $2 trillion in 2017, with IDC projecting the amount spent to reach $772.5 billion in 2018, 14.6% more than the $674 billion it estimated was spent in 2017, with it hitting $1 trillion in 2020 and $1.1 trillion in 2021[1].

---

1 IoT Devices: https://internetofthingsagenda.techtarget.com/definition/IoT-device

This uncontrollable explosion in smart devices will pave the way towards ubiquitous services in different application domains (i.e. healthcare, smart cities, energy management, military, environmental monitoring, and industry automation, to name a few (Palattella et al., 2012)). Moreover, it further motivates the current efforts of the research community, industry, and standardization bodies worldwide devoted to solving a plethora of issues still open in this ebullient context (Atzori et al., 2010a; Miorandi et al., 2012; Palattella et al., 2012). The smart objects forming and enabling the IoT are generally resource constrained (i.e. limited computational, storage, and communication capabilities). They usually communicate over loosely coupled short-range low-power wireless links up to a centralized storage and computational element, able to provide coordination capabilities for the entire network.

In recent years, the industrial IoT (IIoT) has appeared as a popular paradigm to completely reshape the automation experience towards the realization of industrial process automation. The IIoT provides a set of standards to integrate a range of manufacturing equipment to a web-based system and ensures the provision of valuable data to expert systems for smart decision making. The IIoT also makes it possible to connect a wide variety of tiny sensor devices installed throughout the production line to the analytic systems and strives to achieve the following four objectives:

- First, monitoring of data generated by sensor devices in a cost effective and energy efficient way considering that sensors are battery powered in nature.
- Second, processing of this data, transforming it into useful information employing some business analytic tools.
- Third, provision of this actionable information to an expert system/personnel for onward decision making.
- Fourth, attaining the anticipated performance improvement by taking preventive or corrective actions based on the actionable information at hand.

IIoT plays its role by inducing the ultimate performance improvement throughout the production process that leads to huge financial savings.

This chapter aims at shedding light on low power technologies as the key enablers for today's industrial automation. It further discusses the different types of costs involved in harsh industrial environments, with a special focus on the damage penalty and battery replenishment costs for the sensor motes deployed across the production line. Moreover, it will strive to analyse the reciprocal relationship and trade-off analysis for these costs to suggest some optimization strategies for ultimate production efficiency.

7.2 The Evolution of Low Energy Networking Protocols for Industrial Automation

Low energy networking protocols (Lee and Hyoseok, 2016; Xiong et al., 2015) are considered to be trendsetters in the evolution of wireless communications thanks to their inherent capability to match radio coverage, scalability, and energy efficiency requirements for IIoT deployment. Although low energy consumption is one of the characteristics of low energy network protocols such as long range wide area networks (LoRaWAN) (Sherazi et al., 2018), IPv6 over low power wireless personal area networks (6LoWPAN) (Al-Kashoash and Kemp, 2016), and Bluetooth low energy (BLE) (Heydon, 2013) to name a few, but energy exhaustive operation of sensor nodes (also known as motes) installed within a harsh environment or inaccessible places (e.g., in environmental and industrial monitoring use cases) makes it impractical for the administration of a smart industry to replenish the batteries frequently. Moreover, these batteries are an expendable resource with adverse environmental effects. This section presents an overview of the aforementioned low energy networking protocols and highlights the key technologies and standards proposed in the recent past.

7.2.1 Radio Frequency Identification and Near Field Communication

Radio frequency identification (RFID) Nath et al. (2006) has been used to track inventory and for supply chain management. The passive tags on a range of products contain product information that can be read via special handheld readers within a distance of 100 m, supporting unidirectional communication. Similarly, near field communication (NFC) Coskun et al. (2013) can also read/write tags but, unlike RFID, these tags can be employed in an unlimited number of products due to their simple design and any regular NFC enabled device can read their information. NFC also supports bi-directional communication and, according to an estimate[2], it has been used in over a billion devices so far.

7.2.2 Bluetooth

The Bluetooth standard Bhagwat (2001) was another game changing invention in the mid 1990s, facilitating the connectivity of a huge number of handheld peripherals with personal computers. It has been the hot choice for a variety of devices to establish short range (often in the range of 10 m) connections with each other and it was assumed to support relatively a higher data rate than its predecessors. Bluetooth can support the pairing of up to eight devices at the cost of higher energy consumption in comparison to other candidates where each pairing device is an independent component.

7.2.3 Zigbee

The Zigbee (Wang et al., 2006) standard, conceived by Zigbee Alliance in 2005, was another success story in the series of short range technologies. It offered to optimize the communication range as well as lowering the energy consumption significantly as compared to Bluetooth. Zigbee devices could support a radio coverage of up to 100 m with the provision of relatively higher data transfer rates.

7.2.4 Bluetooth Low Energy (BLE)

Soon after the appearance of Zigbee, the Bluetooth special interest group started working in the same dimensions to overcome the energy issues of Bluetooth technology and, just a year after the inception of Zigbee in 2006, they were able to introduce BLE (Heydon, 2013), which aimed to reduce the energy consumption by putting the connections on sleep mode once inactivity was detected for any timespan.

---

2 https://nfc-forum.org/nfc-bluetooth-and-rfid-unraveling-the-wireless-connections

7.2.5 Wi-Fi

Wi-Fi (Al-Alawi, 2006) is another success story, owned and maintained by the Wi-Fi Alliance, which was introduced as a part of the wireless local area network (WLAN) following the 802.11x family of standards. Wi-Fi can be seen as the most famous and widely used option so far, which is great for providing high speed connectivity (e.g, watching a movie online and downloading a video game) for home and office users (i.e. local area) while bearing a relatively lower cost (i.e. as monthly subscription fee or modem charges). The only downside is the relatively higher power consumption and the provision of this connectivity for a very short vicinity, and the signals might face severe attenuation where long distances are involved between the access point (AP) and the end-user. Several follow up versions of the 802.11x family have continuously been introduced so far, overcoming plenty of issues related to the performance of Wi-Fi.

7.2.6 IPv6 Over Low Power Wireless Personal Area Networks (6LoWPAN)

6LoWPAN (Al-Kashoash and Kemp, 2016) is also considered a key player in supporting a range of industrial applications in the domain of long range and low power networks. Unlike other low power wide area (LPWA) options, 6LoWPAN has a proper network layer in its model with IPv6 addressing capabilities over conventional IEEE 802.15.4. Each of the host nodes is assigned an IPv6 address for end-to-end communication like in conventional IP networks. It can support operations on the unlicensed industrial, scientific and medical (ISM) portion of the radio spectrum with 2.4 GHz and sub-GHz frequency bands offering 16 and 8 channels, respectively. Despite achieving a fair receiver sensitivity of up to −92 dBm while operating in the sub-GHz band, it can offer a radio coverage of up to 100 m and the batteries of 6LoWPAN hosts can generally last around a couple of years after employing different energy optimization techniques (Al-Kashoash and Kemp, 2016).

7.2.7 Low Power Wide Area Networks (LPWAN)

LPWAN (Lee and Hyoseok, 2016; Raza et al., 2017) are one of the few families that came out to serve the dense IoT networks. The LPWAN paradigm can be seen as the key enabler to affordable, efficient, and globally available IoT infrastructures. LPWAN employs existing radio networks to connect endpoints located far away from one another in addition to affordable cost and very low power consumption. Transmissions employing LPWAN can be seen as being many times cheaper than that of cellular networks (e.g., General Packet Radio Services (GPRS), 3rd generation (3G) systems). A fairly high receiver sensitivity achievable through LPWAN permits achieving longer coverage, up to several kilometers, even with the lowest output power consumption. In addition, the types of integrated circuits (ICs) used in this kind of networks are not so expensive (i.e. up to only a few euros (Pulpito et al., 2018)).

LPWAN emerged as a prominent solution for low bit-rate machine to machine (M2M) communication with least cost and power requirements, than a conventional mobile network, to target larger coverage areas. This kind of technological shift was inevitable for battery powered sensors, actuators, and microprocessors to operate efficiently and effectively in a resource constrained environment like IoT. LPWAN are designed with the intention of ensuring smooth operation of these IoT devices for a longer period over single-hop long range communication links among the bulk of IoT devices.

Although local area technologies (as shown in Figure 7.2 and discussed throughout this section) are able to well serve different industrial use cases with short range requirements, they do not fit well for many energy critical applications demanding long radio coverage Auer et al. (2011). On the other hand, although cellular standards are good to cover long geographical areas with higher data rate connectivity, they have been declared unfit for the bulk of IoT use cases because of the huge number of tiny devices that solely rely on their batteries for communication and frequent battery replenishment is not possible.

The cost factor is another significant aspect instigating the need for LPWAN technologies as the conventional cellular modules are not affordable when it comes to dense IoT networks. As the IoT modules belonging to the LPWAN family are considered simple and lightweight devices costing only a few euros Pulpito et al. (2018), and are capable of surviving for a longer period of time (up to ten years), they are being considered the future of the IoT concept. Hence, mobile network operators (MNOs) are putting their full attention to

Comparison Wireless technologies
Peak Data Rate vs Maximum Range

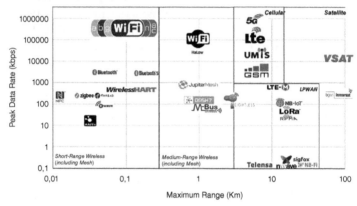

Figure 7.2 Comparison of wireless technologies with respect to maximum radio coverage and data rate support available for industrial applications.

making it happen and they are ready for different public and private network partnerships to provide cheaper low rate services over longer distances (i.e. remote areas), as shown in Figure 7.2. The following are the few popular and widely considered options in LPWAN for a variety of industrial applications.

7.2.7.1 Long Range Wide Area Networks (LoRaWAN)

Long range (LoRa) is emerging as a key technology enabler for low power and long range communications (Augustin et al., 2016). At the beginning, it was particularly intended for low data rate applications developed by a French company Cycleo, later acquired by Semtech. Consequently, it was adopted by the LoRaWAN architecture, i.e. an open-source standard built on top of the proprietary LoRa physical layer (Sornin et al., 2015). At the time of writing, LoRaWAN is a step ahead among other proprietary LPWAN competitors, right from its inception in terms of its widespread deployment for public networks in European market space, openness of the LoRaWAN standard, and flexibility in the choice of available data rates. The first comprehensive LoRaWAN specification 1.0 was released back in 2015 (Sornin et al., 2015). Moreover, a task group under the LoRa Alliance is constantly working towards LoRaWAN 1.1 for conceiving more exciting features (e.g., a roaming and handover mechanism) expected in coming versions (Sornin, 2017). LoRaWAN is now ready to support several use cases of future IoT services in smart cities, smart homes and buildings, smart environments, smart metering, smart agriculture, and many other domains (Semtech Inc., 2017).

At the physical layer, LoRa supports the choice of a flexible number of channels, bandwidth, spreading factors, and code rates to be used for data transmission (Adelantado et al., 2017). The number of channels and their available bandwidth options depend on the target region and the choice of a LoRa vendor. For instance, up to 10 channels can be used in Europe (commonly 3, 6 or 8 channels are selected in real deployments (Bankov et al., 2016)) and 64 in North America. Nodes may transmit in different sub-GHz portions of the spectrum, HF ISM 868 in Europe and 915 MHz in America, with available bandwidth options of 125 kHz/250 kHz/500 kHz.

The spreading factor (SF) is defined as the logarithmic ratio of symbol rate R_s and chip rate R_c. Typical values span from 7 to 12 and the choice of a given spreading factor provides a trade-off between the data rate and communication range. At the same time, SF allows achieving concurrent communications between several end-nodes and a gateway, while preventing interference phenomena. This is true even if the same channel is selected. The code rate is the ratio of the forward error correction to the original data stream to be encapsulated. It is chosen from the range of 4/5 to 4/8. It is important to note that LoRaWAN networks may also employ an adaptive data rate scheme if explicitly requested by the end-node to individually manage the data rate and RF output for that end-node. It aims to optimize the lifetime of end-node batteries and the overall network capacity. Alternatively, end-nodes are free to choose any available channel at any given time and available data rate as their default, by means of pseudo-random channel hopping.

7.2.7.2 Sigfox

Sigfox is another widely considered LPWA option adapted by several MNOs for public network infrastructure. Sigfox is a proprietary solution provider following the top-down approach in the sense that they own the entire network; from all the back-end data and cloud servers to end-node software, but they are open with respect to the end-node's radio technology. They agree to share their radio technology to any silicon manufacturer provided that they agree to abide by their basic terms of services. A lot of manufacturers (e.g., Taxas Instruments, Atmel, STMicroelectronics, and many others) are already on their way to making Sigfox end-nodes. Sigfox strongly believes that keeping the application cost low is the key to driving business (Pham et al., 2017).

Sigfox is an ultra-narrowband (UNB) technology that employs as short as 100 Hz width of band, thus increasing the number of end-nodes per unit bandwidth and further reducing the noise level (Anteur et al., 2015). The Sigfox nodes can achieve up to 100 bps so delay insensitive applications having low bit-rate requirements can be better served by Sigfox. Sigfox can support up to 140 messages of 12 bytes at the uplink and only 4 messages, of 8 bytes each, per day, on the downlink, which is the maximum band occupancy limit imposed by European duty-cycle regulations for the 863–870 MHz band of the spectrum (Nolan et al., 2016).

As Sigfox operates in an unlicensed ISM spectrum like LoRa employing asynchronous communication protocols and can be characterized by inter-channel interference and multi-path fading, and such standards may not guarantee the same quality of service (QoS) as compared to the ones operated in a licensed spectrum (Bardyn et al., 2016). As the application payload size in Sigfox is 12 bytes at maximum this may prove to be a limitation for various IoT applications with higher data rate needs. Moreover, Sigfox suffers in that it has not take much consideration with respect to security, but the team is working on it to overcome this issue soon.

7.2.7.3 Narrowband IoT (NB-IoT)

NB-IoT is a recently introduced standard by the 3rd Generation Partnership Project (3GPP) in Release 13 for a range of IoT applications including smart industries. It employs a subset of the Long Term Evolution (LTE) standard using sub-bands of the LTE network but limits the bandwidth to 180–200 kHz to support a bulk of IoT devices. It is the favorable choice for those MNOs who are interested in upgrading their existing LTE networks to be tuned for M2M communication instead of deployment from scratch. NB-IoT is well capable of guaranteeing QoS requirements operating in a licensed spectrum as compared to other counterparts (Bardyn et al., 2016).

NB-IoT uses orthogonal frequency division multiplexing (OFDM) modulation to achieve a data rate of up to 250 kbps. It can support a payload size of 1500–1600 bytes and can guarantee a link budget of up to 164 dB (Bardyn et al., 2016). The supported data rate in NB-IoT is higher as compared to other proprietary technologies of this family. The NB-IoT also benefits from LTE encryption fulfilling the demands of several use cases and they are planning to come up with an end-to-end security mechanism like LoRa.

NB-IoT employs LTE based synchronous communication protocols that are, of course, optimal to ensure QoS but at the cost of being more expensive. The NB-IoT end-nodes consume more energy (i.e., 120-0300 mAh) as compared to other competitors to meet these QoS requirements. The deployment costs for setting up this kind of network from scratch, and the license costs, are much higher as compared to other LPWAN options.

7.3   An Overview of the Costs Involved in Industry 4.0

Both industry and academia are equally convinced to explore a range of suitable low power protocols and standards keeping in view the challenges laid down by industrial applications. Although extremely low energy consumption is one of the basic characteristics of low power networking protocols, most of the end-nodes employed in this kind of infrastructure are still battery powered in nature. Despite several low power technologies recently being floated to cater to IIoT use cases, energy is still one of the major challenges for this kind of application. It becomes particularly critical in the cases where the associated costs of the manufactured products are significantly higher and timely detection of different types of anomalies at various stages of the production line can work towards avoiding huge financial losses for a smart industry. Achieving this milestone, simultaneously, involves a narrow line trade-off between energy optimization and continuous monitoring during the production process.

Continuous sensing and data collection can obviously be beneficial for generating most updated alerts at one side, but it would cause frequent battery replenishment on the other side. Similarly, a fair sensing interval to generate alerts can well avoid the fast battery drainage hence prolonging the lifetime of sensor nodes, but, sometimes, even a slight latency in popping-up an urgent alert costs a bulk of damaged products, wasting useful resources at the production line. Facing the critical interplay between latency, battery life-time, and the requirements of industry 4.0 applications requires a novel design methodology that accounts for all the costs and benefits entailed by deploying low power protocols in smart industrial environments. There are the following two fundamental operating costs in industry 4.0.

7.3.1 Battery Replacement Cost

Battery replacement cost is the cost incurred on replacing the already drained batteries from the sensor devices installed throughout a smart industry. Despite their fairly long life-time, battery replenishment is inevitable to maintain the ongoing operations in industry 4.0. It becomes even more critical in some industrial applications when the sensor nodes are deployed at a harsh location (e.g., installed within machinery) where it is difficult or almost impossible to substitute the batteries frequently. On the other hand, the core design consideration for battery-powered sensor nodes is the energy efficient operation achieved by tuning the duty cycle, which yields longer battery life at the cost of limiting the performance of the overall system.

7.3.2 Damage Penalty

The damage penalty is the cost of damaged products on the production line during the damage interval in case of an anomaly in the production process. The damage interval is the time between the occurrence and detection of an anomaly during production. An expert system can only issue the alerts immediately after the anomaly detection. Therefore, a latency recorded between the occurrence and the detection of an abnormal behaviour at the production line is extremely critical and can only be reduced by frequent sensing and transmissions. The shorter the sleep interval of the end devices, the less the damage penalty.

7.3.3 Cost Relationships and Trade-off Analysis

Both the aforementioned costs (i.e., battery replacement cost and damage penalty) critically influence the industrial revenue. These are two contradictory costs where, an attempt to reduce one type would cause increasing the other type of cost. They also depend on the type of underlying industrial application and the type of smart industry. For example, damage penalty would be dominant in an industry (such as automotive) where the unit cost of production is fairly high and even a latency of few seconds in detecting abnormal behaviour at the production line can cause ending up with a loss of millions of dollars. On the other hand, the battery replacement cost might be higher than the damage penalty in industries where the unit cost of manufacturing is very low (e.g., in ballpoint production) but the end-nodes still wake up very often (e.g., every minute) in an attempt to timely detect the anomaly. In those case, the sensing and transmission could follow a fairly long interval to optimize the energy consumption to end up with minor battery replacement cost.

To circumvent these issues, cost optimization can dramatically play its role to improve both performance and production efficiency in an industry 4.0. At one end, it significantly reduces the extra operational expenditure (OPEX) incurred by the frequent replacement of batteries, which in turns saves a lot of time as well. On the other hand, it also improves the production efficiency that can be achieved through non-stop industrial operation. It may serve a variety of applications in industry 4.0 to measure and control the behaviour of several use cases, including machine auto diagnosis, water level and temperature monitoring, ozone presence, indoor air quality, ultraviolent radiations, leakage detection, smart production management, and several pollution control systems with the capability of explosive and hazardous gases detection.

7.4   Evaluating Costs in an Industrial Environment: A LoRaWAN Case study

The cost evaluation model envisaged in this chapter considers an implementation of LoRa based monitoring devices in the industrial environment. However, the methodology adopted hereby can also be extended to other low power networking protocols with some customization. Being a part of the IIoT, LoRa end-devices monitor several industrial parameters (such as pollution monitoring, fire detection, flow level monitoring, leakage detection, and temperature monitoring to name a few). The LoRaWAN configuration

Table 7.1 LoRaWAN assumed parameters for the life-time and cost evaluation for industrial automation.

LoRaWAN parameters	Values
Application payload size	1–3 bytes
Modulation method	LoRa (based on chirp spread spectrum)
Spreading factor (SF)	7–12
Coding rate	4/5
Bandwidth	125 kHz
Number of preamble symbols	8
Transmit power	14 dBm

settings considered in the lifetime evaluation are presented in Table 7.1.

It is pertinent to note that an average energy consumption reading for different LoRa SFs is considered in lifetime evaluation. Furthermore, no variation in the energy consumption is observed until the application payload size of 3 bytes, which seems appropriate to several industrial applications for reporting an alert to the expert systems (Sherazi et al., 2018). Being battery powered, LoRa end-devices conventionally follow periodic transmission intervals to prolong their operations. The monitoring devices are put on sleep after each transmission interval until their next measurement. Let the pause time between two consecutive monitoring slots of an LoRaWAN device be the sensing interval, $\mathcal{T}_{sense}$, and t_{sleep} be the duration when the end-device remains in sleep mode then, the sensing interval can be expressed as:

$$\mathcal{T}_{sense} = t_{sleep} + t_{setup} \tag{7.1}$$

where, t_{setup} is the time required for a monitoring node to switch between active and sleep modes (i.e., twice t_{switch}). The sensing interval is critical to expert systems towards timely decision making. At the one hand, a short sensing interval helps detect the anomaly at an early stage and enhances the productivity of a smart industry yielding more revenues. On the other hand, an increased transmission duty cycle of monitoring devices causes short battery lifetime hence frequent battery replenishment is needed. Similarly, where a long sensing interval makes it possible for monitoring devices to maintain their operation for several years, it may simultaneously incur delays in fault detection scenarios hence production efficiency is at stake.

7.4.1 Battery Lifetime of Monitoring Nodes

Several sensing intervals have been considered ranging from one to five minutes to investigate the impact of varying T_{sense} on the energy consumption of LoRa monitoring devices. To this end, LoRa monitoring devices from Semtech Inc. are studied considering the current draw of their chipset as 44 mA when transmitting with 14 dBm output power (Semtech Inc., 2017). The monitoring devices are assumed to be the periodic transmitters only in a unidirectional way. They conventionally follow active and sleep modes where the average current drawn in sleep and switch mode (I_{sleep} and I_{switch}) are 100 nA and 21.9 mA, respectively (Semtech Inc., 2017).

Then, the average charge, Q, in each state (active, sleep, and switch) can be evaluated considering current draws of Semtech's LoRa monitoring node operating in the aforementioned modes and the time duration when a node remains in that state. For example,

$$Q_{TX} = I_{TX} \cdot t_{TX} \tag{7.2}$$

where, I_{TX} is the average current draw of the monitoring node in transmit mode and t_{TX} is the duration of the active period. The total charge, Q_{tot}, is the summation of the products of average current flow and the time duration in all possible states and can be seen as:

$$Q_{tot} = \sum_s I_s \cdot \Delta t_s, s \in TX, \text{sleep, switch.} \tag{7.3}$$

As the total mean energy, $\mathcal{E}_{tot}$, is the product of the total average charge and the voltage applied on the SX1272 end-device so it can be evaluated as:

$$\mathcal{E}_{tot} = Q_{tot} \cdot V_{SX1272}. \tag{7.4}$$

Likewise, the average energy consumed per day, $\mathcal{E}_{day}$, and the average energy consumption during a whole year can also be calculated using Equation 7.4 as follows:

$$\mathcal{E}_{year} = Q_{year} \cdot V_{SX1272}. \tag{7.5}$$

Now, the battery life (in years), $\mathcal{L}_{batt}$, can be evaluated assuming the total battery capacity, C_{batt}, of 1000 mAh (i.e. 11880J @ 3.3V) and can be expressed as follows:

$$\mathcal{L}_{batt} = \frac{C_{batt}}{\mathcal{E}_{day}} \cdot 365. \tag{7.6}$$

7.4.2 Battery Replacement Cost

The battery replacement cost of LoRa monitoring nodes deployed throughout the production line is the first significant cost that is considered critically while designing the monitoring and control system. This cost can further be split into three types of costs: battery purchase cost, installation cost, and the disposal cost for the old battery. The first one is taken as the fixed cost neglecting the inflation factor with time while the second one depends upon the type of industry and the replacement complexity of the LoRa monitoring node whose battery needs replenishment. For example, the monitoring nodes installed internally inside machinery would incur more labour cost due to the complexity of task as compared to the one installed at a relatively simpler spot. Like the purchase cost, the disposal cost can also be assumed as a fixed cost. A set of assumptions followed to evaluate the associated battery replacement cost is presented in Table 7.2.

The battery purchase cost, C_{batt}, can be represented as the total cost of purchasing the required number of batteries in a time period and can be evaluated as:

Table 7.2 Assumptions drawn for evaluating battery replacement cost.

Cost parameters	Assumed values
Mean battery lifetime (in years) evaluated for LoRa device at 14 dBm transmitting power	0.10–5.14
Assumed purchase cost per battery (£), C_b	3.7
Time period (years) considered, T	20
Number of batteries installed per node	1
Variable installation cost per node (£), C_r	3.5–10
Cost for the disposal of batteries in T time period (£), C_{dis}	0.10

$$C_{batt} = C_b \cdot N_{cyc} \tag{7.7}$$

where C_b and N_{cyc} are the cost incurred to purchase a single battery and the number of replacement cycles required in a given time period, respectively. Here, it is important to note that a time period of 20 years is considered for this cost evaluation because it is believed to be the fair lifetime achievable through LoRa monitoring nodes in energy harvesting scenarios (Sherazi et al., 2018). Similarly, the cumulative installation cost, C_{inst}, is the variable labor cost that can be calculated as:

$$C_{inst} = C_r \cdot N_{cyc} \tag{7.8}$$

where C_r is the variable replacement labour cost per node depending upon the complexity of the spot. The battery disposal cost, C_{dis}, is the cost incurred on disposing of the replaced batteries that is not usually higher but it may still be significant in case of large-scale network deployment where thousands of nodes need replacement in a time period. C_{dis} is calculated assuming £1400 as the average disposal cost per ton of battery wastage from the recent statistics published by UK Government authorities (DEF). Hence, the total battery replacement cost in pounds, C_{BRC}, can be seen as the summation of these costs in a time period of twenty years for LoRa monitoring node in the system, and can be expressed as:

$$C_{BRC} = \sum_s C_s, s \in \{batt, inst, dis\} \tag{7.9}$$

7.4.3 Damage Penalty

The damage penalty can be referred as the cost of damaged products on the production line due to a possible delay in anomaly detection. This delay can be seen as the damage interval, T_{dam} and expressed as:

$$T_{dam} = t_d - t_o \; ; \; 0 \leqslant T_{dam} \leqslant T_{sense} \tag{7.10}$$

where t_d and t_o are the anomaly detection time and anomaly occurrence time, respectively. Let P_{dam} be the damage penalty and R_{prod} be the rate of production at the production line (taken in terms of the number of products manufactured per minute), then the damage penalty can be calculated as:

$$P_{dam} = T_{dam} \times R_{prod} \times C_{unit} \tag{7.11}$$

Table 7.3 Product categories considered along with associated unit costs and production rates in different smart industries.

Product category	Unit cost of production, C_u (£)	Rate of Production, R_p (products/min)
Cheap	10	30
Medium	70	6
Expensive	150	3
Very expensive	500	1

where C_{unit} is the unit cost of production assumed for a specific unfinished product. As the domain of damage interval is increased with increasing value of T_{sense}, the damage penalty also keeps on increasing. The damage penalty is evaluated for different categories of product like cheap, medium, expensive, and very expensive, as shown in Table 7.3 .

7.5 Cost Analysis for Industrial Automation

This section spans the results of LoRaWAN evaluation following the proposed model (elaborated in Section 7.4) along with a detailed discussion on these results. The sub-section encompasses the results of several critical performance indicators (such as energy consumption, battery life, spreading factor support, battery replacement cost, damage penalty, and total cost) while evaluating LoRaWAN deployed in industry 4.0.

7.5.1 Statistics for Energy Consumption

Energy consumption can be seen as the foremost LoRaWAN parameter evaluated in the industrial environment. It serves as the primary step to evaluate the battery life of a LoRa monitoring node and it can be evaluated as a product of total charge consumed and the applied voltage by Equation (7.5). Figure 7.3 presents the average energy consumption of LoRa monitoring node per day against a range of fair sensing intervals. The average energy consumption is the average value of all the energy consumptions reported while operating on different

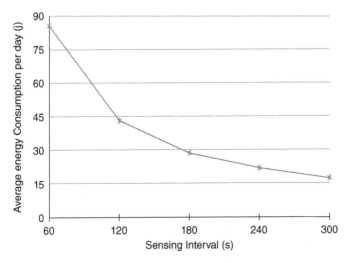

Figure 7.3 Average energy consumption per day against different sensing intervals.

LoRaWAN spreading factors as each of the spreading factors leads to different active and sleep times depending on its air time.

Figure 7.3 reports the maximum value of energy consumption (almost 85 J per day) when the node senses every minute. It is obvious to note that the average consumption goes on decreasing as sensing interval is increased. For example, the average value of energy consumption per day is at a minimum when LoRa monitoring node senses and reports for an anomaly every five minutes. It can be seen as the best case scenario with respect to longer battery life in standard LoRaWAN.

7.5.2 Statistics for Battery Replacement Cost

The battery replacement cost is the first type of cost that an industry may consider significant to increase the revenue. It includes the cost of purchasing the new battery, its replacement labor cost and the cost incurred to properly dispose of the used batteries. The replacement labor cost can be variable depending upon the complexity of the spot where the LoRa monitoring node is installed while the other two costs can be fixed ignoring inflation rate with time.

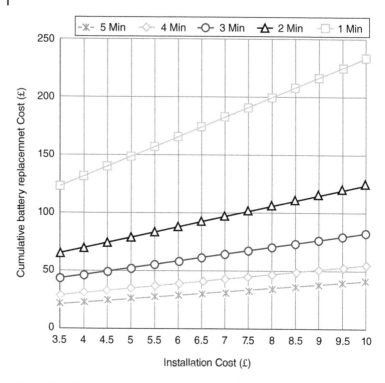

Figure 7.4 Cumulative battery replacement cost against variation with respect to complexity.

Figure 7.4 presents cumulative battery replacement cost when the variation in the installation cost is considered in the range £3.5 to £10 per replacement with respect to the complexity of the spot where the monitoring node is installed within a smart industry. Figure 7.4 clearly explains the linear relationship with increased installation cost. Here, the cumulative replacement cost is much higher for shorter sensing intervals like 1 min. As we move from lower to higher sensing intervals, the cumulative cost increases slightly against the 5 min interval case where the replacement cost is the minimum in a plain industrial environment.

It is obvious from Figure 7.4 that the cumulative battery cost goes on increasing when we go on shrinking the sensing interval because more battery replacement cycles are needed in a 20 year time period when the monitoring nodes wake up more frequently (e.g., in the case of a

1 min sensing interval). Similarly, the replacement cost variation is not affected much when the sensing interval is as longer as 5 min and it does not go beyond £50. The gap between each pair of adjacent curves goes on widening despite an equal increase in the sensing interval, which argues an exponential increment in battery replacement cost with respect to sensing interval.

7.5.3 Statistics for Damage Penalty in a Plain Industrial Environment

The damage penalty can be seen as the second type of cost but high enough to be paid significant attention by the administration of a smart industry. It refers to the cost of damaged products on the production line due to an anomalous production process. Figure 7.5a presents an overview of the damaged penalty against the damage interval when the unit cost is £500 and the rate of production is 1 per min. It can be observed from Figure 7.5a that the penalty goes on doubling when damage interval is increased every minute, starting from 1 min to a maximum of 5 min. It remains unchanged for the duration of each minute until entering into the next minute and keeps on increasing linearly with time. This implies that reducing the damage interval would end up lowering the damage penalty but the damage interval would always span within the range of the sensing interval.

The longer the sensing interval, the longer the damage interval it may cause as smart systems can only detect anomalies to take corrective actions when they first hear the LoRa monitoring nodes after the anomaly occurrence. The best case can be the minimum value of sensing interval so to avoid any delays in detecting the anomalous situation. Similarly, the worst case may be the longest sensing interval when the anomaly occurred just after the previous cycle when last reported by the monitoring node and the smart system would be able to detect this anomaly in the next cycle at the earliest after waiting for the whole sensing interval (e.g., $\mathcal{T}_{sense}$ = 5 min).

Figure 7.5b compares four different product lines from industry 4.0 with different unit costs and the production rates given in Table 7.3. As discussed earlier, all four cases depict the same trend for the damage penalty while moving along sensing intervals. Although there is no noticeable difference between the damage penalty of all four cases

(a)

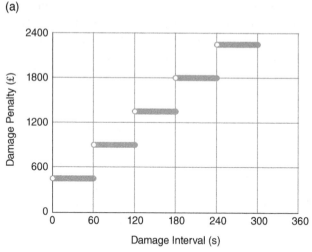

Damage Penalty overview with Damage Intervals

(b)

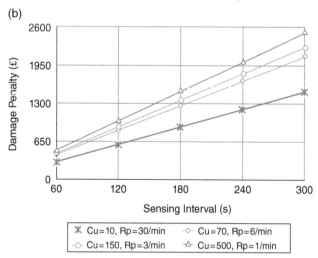

Damage Penalty with respect to various Sensing Intervals

Figure 7.5 Damage interval and penalty demonstration when C_u =£450, $R_p = 1/$min.

on the lower part of sensing interval, as we move on increasing the sensing interval the difference appears to be significant. The product with minimum unit cost with higher production rate seems to be the most ideal case when the penalty does not go beyond £1500, even with the longest sensing interval (i.e. 5 min). The damage penalty may go as higher as £2500 in the case of maximum unit cost and lowest production rate following the same sensing interval.

7.5.4 The Cumulative Cost

The overall cost includes both types of contradictory costs evaluated previously; battery costs and damage penalty. Figure 7.6 throws light on an overall picture depicting both types of cost to estimate a clear contribution of each type of cost in the overall cost. To present an example,the damage penalty is recorded when the unit cost of production is £10 and the rate of production reaches 30 products/min. Both types of costs are comparable with each other in the start when sensing interval is around 1 min.

It is easy to understand from Figure 7.6 that the overall cost goes on increasing with an increase in sensing interval but the proportion of both costs keeps on changing against each other. Initially, the proportion of battery replacement cost is 44% in comparison to damage penalty, which goes down to 3% of the overall cost when

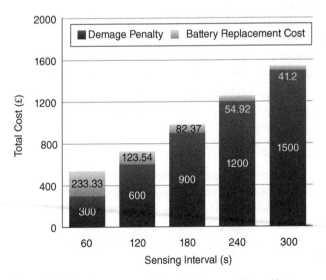

Figure 7.6 Total cost summarizing damage penalty and battery replacement cost when C_u =£10, R_p = 30/min.

the LoRa monitoring nodes reaches 5 min of sensing interval. On the other hand, the damage penalty is doubled after every minute of increment to the sensing interval starting from £300 (when the sensing interval is 1 min) to £1500 if the sensing interval is stretched to 5 min.

7.6 Cost Optimization through Energy Harvesting in Industrial Automation

Energy harvested from renewable energy sources, available in the industrial environment, can dramatically play its role to improve both performance and production efficiency in industry 4.0. In this section, only a few harvesting sources available in the industry are considered to demonstrate the huge cost savings and the great impact of the availability of this surplus energy on the industrial cost optimization process. The following can be the three harvesting sources among an exhaustive list of harvesting sources along with an average energy potential exploitable per day.

- Photoelectric with an average harvestable potential of 4.3 J scavenged during eight working hours every day on 200 lx.
- Thermoelectric with a harvesting potential equivalent to 6.2 J per day for 10 h at 5 °C and for 5 h at 10 °C.
- RF energy with a mean potential of 1.8 J per day with 3 W power transmitted through a 5 m distant source at 9 MHz frequency.

Let $\mathcal{E}^h_{day}$ be the average amount of harvested energy per day available to the industrial environment, the new energy demand per day, $\mathcal{E}'_{day}$, can be evaluated as follows:

$$\mathcal{E}'_{day} = \mathcal{E}_{day} - \mathcal{E}^h_{day} \tag{7.12}$$

where $\mathcal{E}_{day}$ is the previous energy demand per day in a plain industrial environment. Similarly, the new sensing interval can also be derived considering the relaxation in the energy quota due to newly harvested energy as compared to the energy demand in a plain industrial setup. The new sensing interval, $\mathcal{T}'_{sense}$, can be derived from the following expression:

$$\mathcal{T}'_{sense} = \mathcal{T}_{sense} - \left[\mathcal{T}_{sense} \cdot \frac{\mathcal{E}^h_{day}}{\mathcal{E}'_{day}} \right]. \tag{7.13}$$

Hence, in the presence of the aforementioned harvesting energy available at hand, the benefits could be extended toward the following two orientations.

7.6.1 Extending the Battery Lifetime

The battery life of monitoring nodes can significantly be prolonged through the presence of harvested energy. The enhanced battery life would directly impact the battery replacement costs because fewer exchange cycles are required in the same time period, T. Figure 7.7a clearly demonstrates the replacement costs incurred in both plain and energy harvesting industrial environments. It is obvious that as we go along the longer sensing intervals, the number of replenishment cycles are increased so it incurs more replacement costs. Moreover, the battery replacement cost in the energy harvesting scenario always remains lower than in the plain industrial scenario.

Similarly, Figure 7.7b presents an analysis for the damage penalty in both energy harvesting and plain industrial environments considering different industries with different unit costs of production and the production rates presented in Table 7.3. It clearly draws a comparison of damage penalty caused in an energy harvesting environment with the plain industrial environment presented previously in Figure 7.5b. It is obvious from Figure 7.7b that the damage penalty can be restricted to an upper bound limit just above £1000 in the presence of the energy harvesting environment in comparison to the plain industrial environment.

7.6.2 Tuning the Sensing Interval

The second major benefit of harvesting energy from industrial environment can be the tuning of sensing interval. As a surplus amount of energy is available to the monitoring nodes, they can opt to sense and transmit more frequently to update an expert system in short intervals. It would be useful in some industrial lines with very high unit costs of production where even a small latency in detecting an anomaly may cause a bulk of damaged products on the faulty production line. Figure 7.8a shows the rate at which the existing sensing intervals can be contracted in the presence of harvested energy. The percentage of interval contraction also keeps on increasing when we keep on moving

(a)

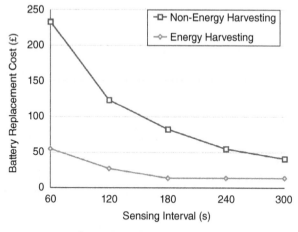

Comparison of battery replacement cost

(b)

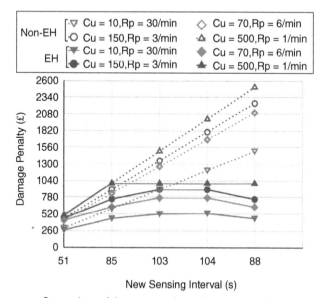

New Sensing Interval (s)

Comparison of damage penalty against new sensing intervals

Figure 7.7 Comparison of battery replacement cost and damage penalty when exploiting energy harvesting sources in an industrial environment.

(a)

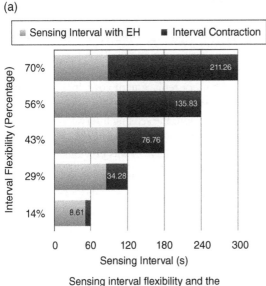

Sensing interval flexibility and the
rate of interval contraction

(b)

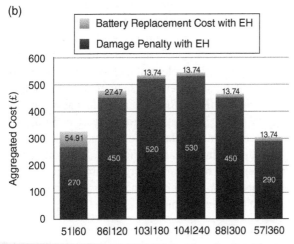

Aggregate cost in an energy harvesting
environment

Figure 7.8 Interval flexibility and aggregate cost optimization achieved
through shortening the sensing intervals when exploiting energy harvesting
sources in an industrial environment.

to the higher sensing intervals. It is evaluated through the relaxation of energy quota when the node was purely battery operated.

Figure 7.8b presents the most significant impact of sensing interval contraction on the aggregate cost in an energy harvesting industrial environment. Here, it can be observed that damage penalty is the dominant cost as compared to battery replacement costs in a harvesting industrial environment. Figure 7.8b argues that the aggregate cost no longer has a linear relationship with the sensing interval as compared to Figure 7.6. It is pertinent to note that the aggregate cost starts declining after the sensing interval of 104 s (which was previously 4 min) to as low as £300 in the case of a 6 min sensing interval.

7.7 Conclusion

This chapter presents a model to evaluate different types of cost associated with industrial monitoring and highlights the cost optimization achieved through a surplus renewable energy harvested within the industrial environment. The chapter throws light on the following critical conclusions. First of all, the damage penalty is the dominating cost in an industrial environment that needs to be cut down with priority. Second, the battery replacement costs is yet another critical cost that could be reduced significantly in a harvesting industrial environment because of the prolonged battery lives of monitoring nodes. Third, harvesting energy enables industrial automation to achieve new heights by optimizing the costs many-fold as compared to the plain industrial setup. Fourth, the linear increase in the aggregate costs is no longer evident thanks to the presence of harvesting sources available within the industrial environment. Fifth, the aggregate cost in the presence of harvesting energy sources can be optimized many-fold, especially for longer sensing intervals, as compared to aggregate costs in a plain environment.

Bibliography

Technical report.

F. Adelantado, X. Vilajosana, P. Tuset-Peiro, B. Martinez, J. Melia-Segui, and T. Watteyne. Understanding the limits of lorawan. *IEEE Communications Magazine*, 55(9):34–40, 2017. ISSN 0163-6804. doi: 10.1109/MCOM.2017.1600613.

Adel Ismail Al-Alawi. Wifi technology: Future market challenges and opportunities. *Journal of computer Science*, 2(1):13–18, 2006.

HAA Al-Kashoash and Andrew H Kemp. Comparison of 6lowpan and lpwan for the internet of things. *Australian Journal of Electrical and Electronics Engineering*, 13(4):268–274, 2016.

Mehdi Anteur, Vincent Deslandes, Nathalie Thomas, and Andre-Luc Beylot. Ultra narrow band technique for low power wide area communications. In *Global Communications Conference (GLOBECOM), 2015 IEEE*, pages 1–6. IEEE, 2015.

L. Atzori, A. Iera, and G. Morabito. The Internet of Things: A survey. *Computer Networks*, 54(15):2787–2805, Oct. 2010a.

Luigi Atzori, Antonio Iera, and Giacomo Morabito. The internet of things: A survey. *Computer Networks*, 54(15):2787–2805, 2010b.

Gunther Auer, Vito Giannini, Claude Desset, Istvan Godor, Per Skillermark, Magnus Olsson, Muhammad Ali Imran, Dario Sabella, Manuel J Gonzalez, Oliver Blume, et al. How much energy is needed to run a wireless network? *IEEE Wireless Communications*, 18(5), 2011.

Aloÿs Augustin, Jiazi Yi, Thomas Clausen, and William Mark Townsley. A Study of LoRa: Long Range & Low Power Networks for the Internet of Things. *Sensors*, 16(9):1466, 2016. ISSN 1424-8220. doi: 10.3390/s16091466.

D. Bankov, E. Khorov, and A. Lyakhov. On the limits of lorawan channel access. In *2016 International Conference on Engineering and Telecommunication (EnT)*, pages 10–14, Nov 2016. doi: 10.1109/EnT.2016.011.

Jean-Paul Bardyn, Thierry Melly, Olivier Seller, and Nicolas Sornin. Iot: The era of lpwan is starting now. In *European Solid-State Circuits Conference, ESSCIRC Conference 2016: 42nd*, pages 25–30. IEEE, 2016.

Pravin Bhagwat. Bluetooth: technology for short-range wireless apps. *IEEE Internet Computing*, 5(3):96–103, 2001.

Vedat Coskun, Busra Ozdenizci, and Kerem Ok. A survey on near field communication (nfc) technology. *Wireless personal communications*, 71(3):2259–2294, 2013.

B. Emmerson. M2M: the Internet of 50 billion devices. *Huawei Win-Win Magazine Journal*, (4):19–22, Jan. 2017.

Ericsson. More than 50 billion connected devices. Ericsson White Paper, Feb. 2018.

D. Evans. The internet of things, how the next evolution of the internet is changing everything. Cisco Internet Business Solutions Group (IBSG). White paper, Apr. 2011.

L. A. Grieco, A. Rizzo, S. Colucci, S. the, G. Piro, D. Di Paola, and G. Boggia. IoT-aided robotics applications: technological implications, target domains and open issues. 54:32–47, December 2014.

Jayavardhana Gubbi, Rajkumar Buyya, Slaven Marusic, and Marimuthu Palaniswami. Internet of things (iot): A vision, architectural elements, and future directions. *Future Generation Computer Systems*, 29(7): 1645–1660, 2013.

O. Hersent, D. Boswarthick, and O. Elloumi. *The Internet of Things: Key Applications and Protocols*. Wiley, 2012.

Robin Heydon. *Bluetooth Low Energy: the Developer's Handbook*, volume 1. Prentice Hall Upper Saddle River, NJ, 2013.

Semtech Inc. Low power long range transceiver, sx1272/73 datasheet, Nov. 2015. URL https://www.semtech.com/uploads/documents/ sx1272.pdf. [Revised Mar. 2017].

Brian Lee and YI Hyoseok. Low power wide area network, November 24 2016. US Patent App. 15/093,969.

Daniele Miorandi, Sabrina Sicari, Francesco De Pellegrini, and Imrich Chlamtac. Internet of Things: Vision, Applications & Research Challenges. *Ad Hoc Networks*, 2012.

Badri Nath, Franklin Reynolds, and Roy Want. Rfid technology and applications. *IEEE Pervasive Computing*, (1):22–24, 2006.

Keith E Nolan, Wael Guibene, and Mark Y Kelly. An evaluation of low power wide area network technologies for the internet of things. In *Wireless Communications and Mobile Computing Conference (IWCMC), 2016 International*, pages 439–444. IEEE, 2016.

M. Palattella, N. Accettura, X. Vilajosana, T. Watteyne, L. Grieco, G. Boggia, and M. Dohler. Standardized Protocol Stack for the Internet of (Important) Things. *IEEE Commun. Surveys Tuts*, 2012.

Congduc Pham, Fabien Ferrero, Mamour Diop, Leonardo Lizzi, Ousmane Dieng, and Ousmane Thiaré. Low-cost antenna technology for lpwan iot in rural applications. In *Advances in Sensors and Interfaces (IWASI), 2017 7th IEEE International Workshop on*, pages 121–126. IEEE, 2017.

M. Pulpito, P. Fornarelli, C. Pomo, P. Boccadoro, and L. A. Grieco. On fast prototyping lorawan: a cheap and open platform for daily experiments. *IET Wireless Sensor Systems*, 8(5):237–245, 2018. ISSN 2043-6386. doi: 10.1049/iet-wss.2018.5046.

U. Raza, P. Kulkarni, and M. Sooriyabandara. Low power wide area networks: An overview. *IEEE Communications Surveys Tutorials*, PP (99):1–19, 2017. ISSN 1553-877X. doi: 10.1109/COMST.2017.2652320.

Semtech Inc. LoRa Use Cases, 2017. URL http://www.semtech.com/ wireless-rf/internet-of-things/lora-applications/briefs. Accessed: 12 April 2017.

H. H. R. Sherazi, M. A. Imran, G. Boggia, and L. A. Grieco. Energy harvesting in lorawan: A cost analysis for the industry 4.0. *IEEE Communications Letters*, 22(11):2358–2361, Nov 2018. ISSN 1089-7798. doi: 10.1109/LCOMM.2018.2869404.

H. H. R. Sherazi, G. Piro, L. A. Grieco and G. Boggia, "When Renewable Energy Meets LoRa: A Feasibility Analysis on Cable-Less Deployments," in *IEEE Internet of Things Journal*, vol. 5, no. 6, pp. 5097–5108, Dec. 2018. doi: 10.1109/JIOT.2018.2839359

N. Sornin. LoRaWAN Specification Development. Technical report, LoRa Alliance, 2017.

N. Sornin, M. Luis, T. Eirich, T. Kramp, and O. Hersent. LoRaWAN Specification v.1.0. pages 1–82, 2015. URL https://www.lora-alliance. org/portals/0/specs/LoRaWAN Specification 1R0.pdf.

Dong Wang, Jin-rong Zhang, Yan WEI, Chang-xiu CAO, and Zheng TANG. Building wireless sensor networks (wsns) by zigbee technology [j]. *Journal of Chongqing University (Natural Science Edition)*, 8:023, 2006.

Rolf H Weber and Romana Weber. *Internet of things*, volume 12. Springer, 2010.

Xiong Xiong, Kan Zheng, Rongtao Xu, Wei Xiang, and Periklis Chatzimisios. Low power wide area machine-to-machine networks: key techniques and prototype. *IEEE Communications Magazine*, 53(9): 64–71, 2015.

8

A Non-Event Based Approach for Non-Intrusive Load Monitoring

Ahmed Zoha, Qammer H. Abbasi, and Muhammad A. Imran

James Watt School of Engineering, University of Glasgow, UK

8.1 Introduction

Research to date has mainly focused on event based approaches to discern appliances from composite load measurements (Ehrhardt-Martinez et al., 2010; Zeifman and Roth, 2011; Zoha et al., 2012a,b). Until, recently, there have been attempts to directly estimate the sources that compose the aggregated signal without detecting specific events from power measurements. However, such preliminary studies were undertaken to devise a completely unsupervised load disaggregation system. Moreover, it is worth noting that without sub-metered training data, the methods did not achieve high enough disaggregation performance and at present cannot be envisioned for a real-world scenario.

This is the exact challenge addressed in this chapter. To do so, we propose a non-event based probabilistic model to monitor the usage of desk level appliances in a non-invasive manner. The aim is to achieve high disaggregation accuracy by first learning the accurate parameters of appliance models using the segregated signatures of each target appliance and later utilize them in recognizing appliance state transitions within the aggregated load measurements.

Wireless Automation as an Enabler for the Next Industrial Revolution, First Edition.
Edited by Muhammad A. Imran, Sajjad Hussain and Qammer H. Abbasi.
© 2020 John Wiley & Sons Ltd. Published 2020 by John Wiley & Sons Ltd.

The problem of non event based appliance disaggregation can be formally expressed as follows. Given the sequence of aggregated power readings $\mathbf{Y} = [y_1, \ldots\ldots y_T]$ of M appliances for $\mathbf{t} = [1, \ldots T]$ time measurements, we want to discern the power contribution of each appliance $\mathbf{p} = [p_1^{(m)} \ldots\ldots p_T^{(m)}]$ where $\mathbf{p}$ is dependent on the states of the appliances $S_t^{(m)} = [s_1^{(m)}, \ldots\ldots s_T^{(m)}]$ such that $m \in \{1, \ldots, M\}$. At any point in time t $Y_t = \sum_{i=1}^{i=M} p_t^i$, whereas the consumption information of each appliance state can be determined from the sub-metered data during the training phase. Hence, the problem is thus reduced to determining the states of the appliances $S_t^{(m)}$ during each time period t.

In an office environment where multiple desk level appliances are connected to a single power outlet, a probabilistic model can be used to perform hidden appliance state estimation given aggregated power observations. Therefore, in this study, we have decided to investigate the suitability of factorial hidden Markov models (FHMMs) for desk level load disaggregation. An important aspect of our work is the selection of adequate feature sets, which are used for the proposed classifiers and corresponding modelling of power states of individual appliances. Through empirical evaluations, we showed that concatenation of power and statistical features can not only improve the binary state (ON/OFF) estimation of appliances, but it works well even for the inference of multi-state power consumption models. Current non-event based load disaggregation studies have mainly focused on recognizing binary load operations; however, in a real-world setting many appliances often operate in more than two states. Considering this, we evaluated the suitability of our models for binary- and multi-state appliance operations. Our proposed solution does not rely on high-fidelity power measurements; instead, we make use of low-frequency power measurements for our target models. Such a solution offers scalability because the existing metering infrastructure, particularly in a residential environment, provides low-resolution data. Moreover, we have shown that our approach works in real time once the models are trained as opposed to traditional approaches, and we present a deployment of our system as a live application.

The probabilistic framework for modelling and estimation of hidden appliance is discussed in the following section.

8.2 Probabilistic Modelling for Load Disaggregation

Hidden Markov models (HMMs) (Rabiner, 1989) have been widely used to model stochastic processes and are also well suited to model a combination of independent processes. A graphical representation of an HMM model is shown in Figure 8.1(a). S_t is a hidden state variable that affects the distribution of the observable state variable Y_t. In the context of load disaggregation, Y_t is the power signal observed at the output of the energy meter governed by S_t, which represent the underlying states of an appliance. For example the LCD screen can have three operational states: ON, IDLE and OFF. The possible state transitions of an LCD screen can be represented as shown in Figure 8.1(b). This can be translated into an HMM model λ by defining initial state probability π, emission probability ϕ and state transition probability A s.t $\lambda = \{\pi, \phi, A\}$. π defines the initial probability of an appliance state at $t = 1$, whereas A is a transition matrix representing the possible state transitions within a model. ϕ is the probability of an observation at time t given a particular state.

In a simplistic scenario, Y_t at time step t can be thought of as power drawn values of an appliance in a particular state k and we assume ϕ to follow a Gaussian distribution: $\phi \approx \mathcal{N}(\mu_k, \sigma_k)$, where μ and σ are the mean and standard deviation of the output observation. The presence of learning algorithms as proposed in (Rabiner, 1989), for training the HMM (e.g. the Baum–Welch algorithm), evaluating the model likelihood (e.g. the forward backward algorithm) as well as for the

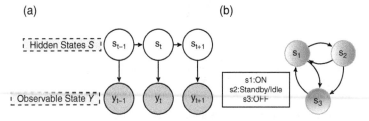

Figure 8.1 (a) Graphical representation of the hidden Markov model (HMM). S_t is a discrete variable whereas Y_t is the observable output at time step t. The vertical arrows show the dependency of the observable output on the hidden state (b) State transition model of an LCD screen.

estimation of probable hidden state sequences (e.g. the Veterbi algorithm) make this model a popular choice for various pattern recognition problems.

Our objective is to perform hidden appliance state estimation given the aggregated power readings. To represent a combined load model for the appliances operating in parallel, it is possible to define a regular HMM model with $K^M \times K^M$ transition matrices, where K is the number of states in each appliance and M is the total number of target appliances. However such a model would impose high computational requirements as state transitions grow exponentially with an inclusion of each new appliance. The FHMM is an extension to HMM that limits the transition matrices to M matrices of size $K \times K$ by introducing a distributed state space architecture. This variant of HMM has been thoroughly studied by Ghahramani and Jordan (Ghahramani and Jordan, 1997), and, as the exact inference with K^M HMM states becomes intractable, hence efficient algorithms for learning the model parameters as well as the possibility of using faster variational approximation to infer the most likely sequence of states for each of the Markov chains has been proposed. In FHMM, independent Markov chains contribute to a single observable output, as shown in Figure 8.2(a). The hidden state S_t is split into M independent factors $S_t^{(m)}$. However, the transition matrix is constrained

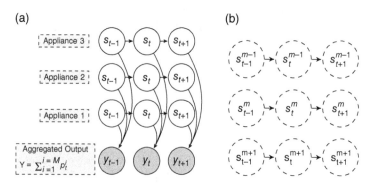

Figure 8.2 (a) Graphical representation of an FHMM, which is a combination of M HMMs. The hidden state is now divided into m factors (i.e. $S_t^{(m)}$). The observable output Y_t is now dependent on contributions from all of the sub-HMMs. The transition matrix is constrained to produce M independent factors (b) A structured variational approximation assumes uncoupling amongst M chains to simplify the inference task.

in a way that there are no intermediate state transitions between the M independent chains but they are still linked via the observable output Y_t.

The arrows denote the horizontal and vertical dependencies between the hidden and the observable states. Each chain or factor m representing one individual appliance can take one K state, $S_t^{(m)} \in \{1, 2, \ldots, K\}$, and the estimated hidden probability distribution is represented as $Q(S_t^{(m)})$. The output contributions from each factor $W_{S_t^{(m)}}^{(m)}$ are dependent on the state of that factor at time t. The observable output on the other hand is the summation of each factor contribution. We further discuss the model definition and provide an overview of learning and inference methods for an FHMM in the following subsections.

8.2.1 Model Definition

As discussed earlier, each chain in the FHMM follows Markovian dynamics, meaning a hidden state at time t (i.e. $S_{t=t}^{(m)}$) is independent of past states (i.e., $S_{t<t}^{(m)}$), given the immediately preceding state, i.e. $S_{t=(t-1)}^{(m)}$). Formally,

$$P(S^{(m)}) = (S_{t=1}^{(m)}) \prod_{t=2}^{T} (S_t^{(m)} | S_{t-1}^{(m)}). \tag{8.1}$$

Secondly, each chain evolves according to its own dynamics, meaning

$$P(S) = \prod_m p(S^{(m)}). \tag{8.2}$$

The observation Y_t is a Gaussian random vector, whose mean is the summation of the output contributions of each factor as expressed in the following equation

$$\mu_t = \sum_{m=1}^{M} W_{S_t^{(m)}}^{(m)}. \tag{8.3}$$

Let $D \times 1$ be the dimension of the observation vector Y_t, then probability density function of the model output can be expressed as

$$P\{Y_t | S_t\} = |C|^{-\frac{1}{2}} (2\pi)^{-\frac{D}{2}} \exp\left\{ \frac{1}{2} (Y_t - \mu_t)^T C^{-\frac{1}{2}} (Y_t - \mu_t) \right\} \tag{8.4}$$

where C is a $D \times D$ covariance matrix and the mean μ_t is dependent on the respective contributions from the appliances at time step t, as expressed in Equation 8.3. As pointed out earlier, the chains are independent of each other, and their horizontal dependencies can be modelled as M transition matrices $A^{(m)}$. This is equivalent to combining independent appliance HMM, and using the properties of Equations 8.1 and 8.2; the hidden state priors and transition matrix for the FHMM can be formally expressed as Equations 8.5 and 8.6.

$$P(S_1) = \prod_{m=1}^{M} \pi^{(m)} \tag{8.5}$$

$$P(S_t|S_{t-1}) = \prod_{m=1}^{M} A^{(m)}. \tag{8.6}$$

This independence means that the probability of appliance 1 being in state a, appliance 2 being in state b and so on, is the product of separate marginal probabilities. Thus, the joint probability distribution $p(Y_t, S_t)$ can be defined as follows

$$P(Y_t, S_t) = P(S_1)p(Y_1|S_1) \prod_{t=2}^{T} P(S_t|S_{t-1})P(Y_t|S_t). \tag{8.7}$$

Equation (8.7) can be expanded using Equation (8.5) and Equation (8.6)

$$P(Y_t, S_t) = \prod_{m=1}^{M} \pi^{(m)} P(Y_1|S_1) \prod_{t=2}^{T} \prod_{m=1}^{M} A^M P(Y_t|S_t). \tag{8.8}$$

8.2.2 Inference

The expectation maximization (EM) algorithm learns the parameters of the HMM in two iterative steps. In the expectation step (E-step), the inference of the posterior distribution of model states $P(S|Y, \lambda)$ is performed whereas in the maximization step (M-step) the parameters of the model λ are updated to their maximum likelihood values. However, in the case of the FHMM, exact inference in the E-step is a computationally expensive process and the probabilities of interest become intractable to compute. To overcome this issue and to decrease the computational requirement several approximate inference methods have been proposed. The author in (Ghahramani and Jordan, 1997) has provided a comparison between the exact and approximate methods for training and inference in the FHMM. In approximate methods, a simplified graph structure is assumed with an introduction of approximate distribution $Q(S)$ with the aim to minimize the Kullback–Leibler (KL) divergence between approximate $Q(S)$ and exact distribution $P(S)$. In this study, we adopt a structured variational EM approximation method, which is briefly discussed below. The justification and complete derivation of the model is provided in (Ghahramani and Jordan, 1997). The simplified structure for the structured variational approximation method is shown in Figure 8.2(b), where it is assumed that M Markov chains are uncoupled. The inference problem is simplified in the E-step by introducing a responsibility factor $h_t^{(m)}$ in place of $P(Y|S)$. $h_t^{(m)}$ can be thought of as a fictitious observation that represents a combination of different settings for $S_t^{(m)}$. The probability of this responsibility factor is varied to minimize the KL divergence between $Q(S)$ and $P(S)$ during the E-step. Hence, the parameters of the approximate distribution becomes $\lambda = \{\pi^{(m)}, A^m, h_m^t\}$. The $Q(S)$ can be written as

$$Q(S|\lambda) = \frac{1}{Z_Q} \prod_{m=1}^{M} Q(S_1^{(m)}|\lambda) \prod_{t=2}^{T} Q(S_t^{(m)}|S_{t-1}^{(m)}, \lambda) \qquad (8.9)$$

where prior and transition probabilities can be written in terms of h_m^t as follows

$$Q(S_1^{(m)}|\lambda) = \prod_{k=1}^{K} (h_{1,k}^{(m)} \pi_k^{(m)})^{S_{1,k}^{(m)}} \qquad (8.10)$$

$$Q(S_t^{(m)}|S_{t-1}^{(m)}, \lambda) = \prod_{k=1}^{K} \left(h_{1,k}^{(m)} \prod_{j=1}^{K} (A_{k,j}^{(m)})^{S_{t-1,j}^{(m)}} \right)^{S_{t,k}^{(m)}}. \qquad (8.11)$$

Comparing Equation 8.9 with Equation 8.8 one can see that the $P(Y_1|S_1)$ is factorized by the observation likelihood corresponding to each factor using $h_t^{(m)}$. Hence, the most likely state sequence within each factor or posterior probability for each chain can be carried out independently as in case of the single HMM, with K states. The M-step for the FHMM is same as for the HMM and is tractable, the details of which can be found in (Ghahramani and Jordan, 1997). A generative approach to learn the parameters of the model is summarized in Algorithm 1.

The process of probabilistic load modelling for appliance disaggregation is summarized as follows: the first step is to initialize the FHMM parameters and this requires selecting the total number of chains in the model that corresponds to the total number of appliances. Moreover, each chain consists of a different number of

---

Algorithm 1 Generative approach to parameter learning.

$\lambda \leftarrow$ Initialize parameters
repeat
$\quad \lambda' \leftarrow \lambda$
$\quad \lambda \leftarrow \arg\max_{\lambda} E\left[log\ Q(Y, S|\lambda)|Y, \lambda')\right]$
until λ converges
$S* = \arg\max_{S*} Q(Y, S|\lambda)$

---

states that depends on the operational behaviour and type of the appliances. To initialize the load disaggregation model, the initial state and transition probability must be specified as explained in Section 8.2.1. The next step is to use the EM algorithm as summarized in Algorithm 1 to learn the model parameters. The EM algorithm iterates in two steps until convergence is achieved. The convergence of the algorithm is measured until the difference of change between the old and new model parameters becomes less than a specified value, which is referred to as the stopping criteria. As discussed earlier in this section, the objective of the structured variational approximation is to minimize the KL divergence as it offers a theoretical assurance that the lower bound on the likelihood is maximized during the E- and M-steps. Each iteration of the E- and M-steps increases the likelihood $Q(Y|\lambda)$, until convergence to the local optimum.

Finally, the goal is to discover the hidden states of the appliances, given the aggregated power measurements. Once the model parameters are learned, the sequence of hidden states variables can be decoded using the maximum likelihood estimation (MLE) principle. We want to find the joint probability of hidden states that has generated the observed signal. The decoding of the most likely sequence of states can be done via applying standard Viterbi algorithm (Rabiner, 1989) that finds the maximum likelihood over all possible state sequences, which can be formally expressed as

$$S* = \arg\max_{S*} Q(Y, S|\lambda) \qquad (8.12)$$

8.3 Experimental Evaluations

The proposed approach has been evaluated by acquiring the data from the experimental setup, as discussed in detail in (Nati et al., 2013). The hardware for the data acquisition module comprises of a smart power outlet (SPO) that acts as a circuit level monitoring device. The SPO consists of a TelosB mote that has been interfaced with an off-the-shelf energy meter (Plogg), and is collocated to a target desk so that a number of appliances can be attached to it via a multi-socket. A subset of appliances including a desktop computer, an LCD screen, a laptop, a fan with three distinct states and an incandescent desk lamp having two states of operation is used in the experiments. The SPO was used to measure the aggregate device usage of appliances. To evaluate the performance of our proposed approach, the experiments were conducted in two phases as discussed below. Implementation and evaluation were performed using a Matlab environment.

8.3.1 Experiment Design

To train and test the performance of our non-event based appliance models, both aggregate and per appliance level data were acquired in two phases: the binary and multi-state operational phases. In the binary phase, we configured all the target appliances to operate just in two states: on and off. For example, using the power-management options we disabled any power-saving settings for the LCD screen (LS), desktop computer (DC) and laptop, so that they would continue to operate in active mode without switching to intermediate states. Conversely, for the multi-state phase, all possible state transitions were taken into account for the target devices. Appliances were individually monitored for an average duration of 30 min during these two phases, and the acquired data was used to derive their respective HMM chains (i.e. the number of states, transition matrix and initial state probabilities) using the EM algorithm. The parameters learned were further combined using Equations 8.5 and 8.6 to define the initial parameters of the composite load FHMM. The FHMM further estimates the model parameters using Algorithm 1. Once the model was trained, the appliance state estimation was performed using the observable power measurements.

To explore the effectiveness of the proposed approach for desk level load disaggregation, 10 test scenarios were designed as shown in Table 8.1 with at least two appliances operating in parallel. The mean duration for each appliance state was set to 5 min. Appliances were switched to their respective states during each scenario and data was accordingly annotated to obtain ground truth labels. The target states

Table 8.1 Test scenarios to evaluate the hidden appliance state estimation performance of FHMMs when appliances are operated in parallel.

Test scenarios	Appliance combinations	Binary phase (target states)	Multi-state phase (target states)
1	DC, LCD	4	5
2	Laptop, lamp	4	5
3	Lamp, fan	4	5
4	Laptop, fan	4	6
5	DC, LCD, lamp	6	7
6	DC, lamp, laptop	6	8
7	DC, LCD, lamp, laptop	8	10
8	DC, LCD, lamp, fan	8	10
9	DC, LCD, laptop, fan	8	11
10	All five appliances	10	13

to estimate are dependent on the number of target appliances used in each scenario. The next sub-sections provide the detail of different models that were considered for hidden appliance state estimation.

8.3.2 Feature Sub-Groups

The appliance-specific power measurements can be characterized by extracting a set of features from the voltage and current signals. Let T donate the length of observation vector Y_t, which in our case was set to be 15, and each measurement can be characterized by a D-dimensional feature vector. We used power based features for state modelling, including average real and reactive power, as well as their respective standard deviations and power factor information. Since, we were mainly looking at the steady state behaviour of appliances that could be seen as a stationary stochastic process, the power features were well suited for characterizing the combination of such processes. They are further divided into sub-groups, whereas each of them is used to train an independent FHMM, as listed in Table 8.2. This allowed us to study the effectiveness of the target features and their combinations for appliance state modelling, which consequently impacts the hidden state estimation accuracy. The models trained

Table 8.2 Features for appliance models.

Model ID	Observation vector (Y_t)	Comments
M_1	$\overline{P}$	Where $\overline{P} = \frac{1}{T}\sum_{t=1}^{T} P_t$
M_2	$\overline{P}, \overline{Q}$	Where $\overline{Q} = \frac{1}{T}\sum_{t=1}^{T} Q_t$
M_3	$\overline{P}, \overline{Q}, PF$	Where $PF = \frac{\overline{P}}{AP}$
M_4	$\overline{P}, \overline{Q}, PF, P_{std}$	Where $P_{std} = \sqrt{\frac{1}{T}\sum_{t=1}^{T}(P_t - \overline{P})^2}$
M_5	$\overline{P}, \overline{Q}, PF, P_{std}, Q_{std}$	Where $Q_{std} = \sqrt{\frac{1}{5}\sum_{t=1}^{T}(Q_t - \overline{Q})^2}$

using these five sub-groups were evaluated against the test scenarios listed in Table 8.1.

8.3.3 Performance Evaluation

An overview of FHMM based decoding of appliance states given the power measurements is shown in Figure 8.3. It consists of three steps: the characterization of acquired measurements by extracting power features; decoding of the appliance states using our trained models; and accessing the accuracy of the hidden state estimation by comparing it against the ground truth data. To evaluate the performance of our models, a macro average F measure (F_1 score) was adopted as a evaluation metric. The experimental results are reported in the following discussion.

8.3.3.1 Binary and Multi-State Classification
In the first phase of the test experiments, as listed in Table 8.1, when only binary operation of the appliances is allowed, the

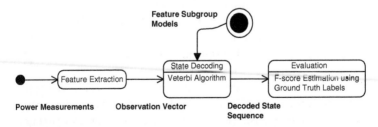

Figure 8.3 An overview of the appliance state decoding process using the feature subgroup FHMMs.

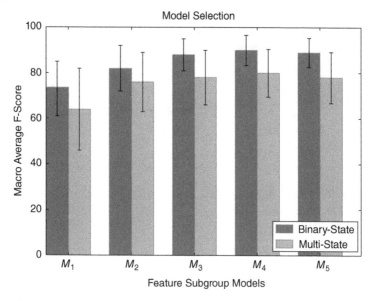

Figure 8.4 Performance comparison of feature subgroup models on target test scenarios.

performance of our target models is summarized in Figure 8.4. It can be seen that model M_4 shows a highest state estimation accuracy of 0.906, whereas the lowest performing model was found out to be M_1 with 0.734 accuracy. The advantage of composite features for modelling the states of the appliances is clear, as model M_2 to M_5 have shown a positive impact on the state decoding performance. However, we did not observe any significant performance improvement for model M_5 compared to M_4. Likewise, in the second phase of the experiment, M_4 still showed superior performance compared to rest of the models, as shown in Figure 8.4. However, it must be noted that a 10% performance degradation was observed during the multi-state appliance operation scenario.

A detailed insight into the best preforming model M_4 reveals some interesting observations. In Table 8.3, we list the performance scores for each test scenario using M_4 for binary and multi-state experimental sessions. One can clearly see that in between test scenarios 7 to 10, the appliance state recognition performance fell below the macro average

Table 8.3 The F scores obtained for the test scenarios using model M_4.

Phase Scenario	1	2	3	4	5	6	7	8	9	10
Binary	0.987	0.980	0.976	0.862	0.941	0.931	0.90	0.88	0.830	0.768
Multi	0.956	0.961	0.87	0.702	0.890	0.91	0.74	0.731	0.669	0.614

score in both experimental phases. These observations led to several findings, summarized as follows:

- Firstly, in comparison to binary-state appliance activations, a clear deterioration in the performance of target models was observed for multi-state appliance operations. This was also true when the number of appliances under test were increased, particularly for cases between 7 and 10. Since, the increase in the number of appliances ultimately led to an increase in the number of appliance states to monitor, the possibility of similar power draw values between appliance intermediate states also increases. This resulted in aliasing of power sums and hence, like other steady-state techniques, in such a situation disambiguation between appliance states becomes a challenge. This problem becomes much worse for generative models like FHMMs, where state decoding of one appliance is dependent on the correct state estimation of another active appliance. For example, the power draw by a laptop and fan running in their active states is equivalent to that of an LCD in an on state. The resulting state estimation error due to this overlap propagates in the chain, thereby leading to high false positives errors while lowering the overall accuracy of the model. This explains the low scores obtained by model M_1, which relied on power information alone. However, one can see that this problem was alleviated by increasing the dimensions of the feature vector.
- Secondly, not only the number, but the type of appliances operating in parallel also had an impact on hidden state estimation accuracy. The combined probability distribution at time step t depends on the interaction of underlying M factors. The state profile of each appliances depends on its internal structure whether it consisted of inductive, capacitive or resistive elements. Reactive power is dependent on the phase angle between the current and voltage, and for capacitive loads the currents leads the voltage and the opposite happens for the inductive loads, thus producing the leading and lagging power factors. Therefore, if loads containing capacitive and inductive elements (capacitors and motors) such as laptop and fans operating in parallel, the reactive powers of the loads, instead of addition, cancel each other out. On the other hand, during the initialization of the FHMM, we assumed that contributions from each factor at time step t linearly combined to represent the observed

variable. This led to inaccurate profiling of probability distribution during the parallel operation of inductive and capacitive loads, and resulted in state estimation errors as seen in case of test scenarios 8 to 10. Resistive loads (i.e. lamps), on the other hand, had no reactive power, so their combination with inductive and capacitive loads did not have any impact on our assumptions. This further explains the fact that model M_5 showed no performance improvement compared to M_4, and the concatenation of Q_{std} with the rest of the power feature for composing a load signature showed no clear advantage. Conversely, during the start-up phase and state transitions appliances showed variability in the power consumption. This was also the case when the fan speed was changed from medium to high. Similarly, the work station load varies depending on the CPU usage. The addition of P_{std} encoded these variations in the composite load signature, which improved the characterization of appliance behaviour in their respective states and the overall estimation accuracy, as evident from the results.

In comparison to composite load operations, our non-event based approach has achieved a much higher accuracy when appliances were operated in a segregated fashion. The appliance recognition accuracy for each individual appliance using the best performing model M_4 is shown in Figure 8.5. The proposed model uniquely profiles the state transition behaviour of each target appliance with more than 90% accuracy, except for the multi-state operation of the desk fan due to its nonlinear behaviour during state transitions.

8.4 Live Deployment

To demonstrate the performance of the non-event based load disaggregation model in a real time, we set up a live deployment that included an office work desk, the target appliances, the SPO unit as well as mobile applications, as shown in Figure 8.6(a). The mobile application was named the 'device workdesk' and was developed to allow a user to monitor the current status of all their desk appliances using a mobile phone. The measurements from the Plogg units were relayed to the server where they were stored in a Mysql database. This database was queried by the monitoring station and the aggregated

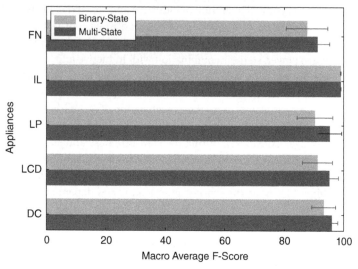

Figure 8.5 Individual load recognition performance using M_4 for binary and multi-state operation scenario.

energy consumption of each work desk was forwarded to the load disaggregation module running in a Matlab environment. The output of the disaggregation module was the predicted states of the target appliances. For the sake of illustration, a combined load profile of a workstation and a LCD monitor are shown in Figure 8.6(b). The load curve was decoded into states by our proposed algorithm. Each power level identifies a combination of states of target appliances, where the consecutive states represent the duration of the state that can be used to estimate the overall power consumption. This information was fed to the mobile application that accordingly displays the current status of the appliances present at the user's work desk.

8.4.1 Energy Estimation

Once the proposed system predicts the underlying appliance states that generates the aggregated signal, it can further estimate the power consumed by each individual appliance. Table 8.4 show the estimated energy consumption of each desk level appliance operated during the

(a)

(b)

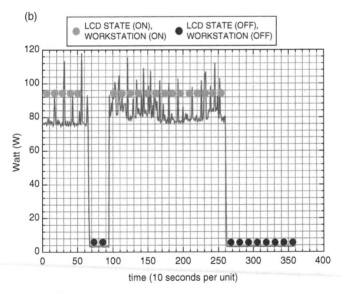

Figure 8.6 (a) Live deployment: mobile application, Plogg unit and the target appliances at a work desk (b) An aggregated load measurement is shown, which is a result of parallel operation of the LS and a workstation. The overlaid green and black dots are the states decoded by our model M_4 and they are forwarded to the mobile application.

Table 8.4 Appliance-level energy consumption estimation.

Appliance	Power-on time (h)	Average power (W)	Energy (W h)	Estimated energy (W h)	Error %
Lamp	6.75	35.6	252.4	240.3	5
LCD	5.3	31.2	178.58	165.36	8
DC	6.2	61.3	437.69	380.6	15
Laptop	5.6	39.2	175.6	219.52	20
Fan	3.2	25.2	58.86	80.64	27

live evaluation of the proposed algorithm. The system simply multiplies the duration of each appliance states with its average power consumption, which was determined during the training phase, to estimate the per appliance energy consumption. The estimated energy for each appliance was then compared to the real energy consumption figures obtained from the Plogg unit. It is to be noted that the estimation error differs for different appliance category. As for the LCD and lamp, the error is less than 10% because of their stable power behaviour; however for the laptop, desktop and fan the energy estimation error is high due to variability in their consumption profile. The misclassification of appliance states also results in an inaccurate energy consumption estimation for these devices.

8.5 Conclusion

In this chapter, we investigated the use of FHMMs to identify the most likely sequences of appliance states that correspond to the time series of aggregated power measurements. It was shown that our proposed non-event based approach could model the aggregated behaviour of desk level appliances with an accuracy of 90% and 80% for binary and multi-state appliance operations, respectively. Moreover, empirical evaluations suggest that non-event based and event based approaches are competitive in performance for recognizing individual appliance operations. However, for composite load operations the performance of the FHMMs were found to be sensitive to the number and type of appliances active in parallel.

Bibliography

Karen Ehrhardt-Martinez, Kat A. Donnelly, and John A. Laitner. Advanced metering initiatives and residential feedback programs: A meta-review for household electricity-saving opportunities. Technical Report E105, American Council for an Energy-Efficient Economy (ACEE), Washington, DC, 2010.

Zoubin Ghahramani and Michael I Jordan. Factorial hidden markov models. *Machine Learning*, 29 (10): 245–273, 1997.

M. Nati, A. Gluhak, H. Abangar, and W. Headley. Smartcampus: A user-centric testbed for internet of things experimentation. In *Accepted for publication to 16th International Symposium on Wireless Personal Multimedia Communications*, 2013.

L R Rabiner. A tutorial on hidden markov models and selected applications in speech recognition. *Proceedings of the IEEE*, 77 (2): 257–286, 1989.

Michael Zeifman and Kurt Roth. Nonintrusive Appliance Load Monitoring : Review and Outlook. *In IEEE Trans. Consum. Electron.*, 57: 76–84, 2011.

A. Zoha, A. Gluhak, M. Nati, M.A. Imran, and S. Rajasegarar. Acoustic and device feature fusion for load recognition. In *In Proceedings of 6th IEEE International Conference on Intelligent Systems (IS'12),Sofia, Bulgaria*, volume 1, pages 386 –392, sept. 2012a. doi: 10.1109/IS.2012.6335166.

Ahmed Zoha, Alexander Gluhak, Muhammad Imran, and Sutharshan Rajasegarar. Non-intrusive load monitoring approaches for disaggregated energy sensing: A survey. *Sensors*, 12 (12): 16838–16866, 2012b.

9

Wireless Networked Control

Zhen Meng and Guodong Zhao

James Watt School of Engineering, University of Glasgow, UK

9.1 Introduction

The development of communication technology has promoted the development of new technologies. On the other hand, some emerging technologies have also put forward higher requirements for communication technologies. The two advances have shown a spiral upward trend. In particular, in recent years, with the development of the interdisciplinary technology pushed by the fourth industrial revolution, the realization of some emerging applications like tactile internet (van den Berg et al., 2017), the Internet of Things (IoT) (Ahlgren et al., 2016), cyber physical systems (CPSs) (Lee, 2008), industrial automation and smart grids (Farhangi, 2010) have been made possible. Based on their time and space distribution characteristics and both real-time performance and highly reliable control requirements it is expected that wireless network control will play a crucial role in the near future.

However, the development of these interdisciplinary technologies often cannot be simply spliced by multiple disciplines of technology but requires multiple parties to infiltrate and integrate with each other. The existing solution of wireless communication control does not yet offer sufficient performance with respect to real-time and reliability requirements and is still not widely used in industrial automation. For example, the tactile internet and industrial automation need to overcome the 1 μs end-to-end latency challenge while the current solution given by cellular network is approximately 15 μs

Wireless Automation as an Enabler for the Next Industrial Revolution, First Edition.
Edited by Muhammad A. Imran, Sajjad Hussain and Qammer H. Abbasi.
© 2020 John Wiley & Sons Ltd. Published 2020 by John Wiley & Sons Ltd.

round-trip latency (Simsek et al., 2016). Smart grids require as low as 10 μs latency with 10^{-9} packet loss probability (Luvisotto et al., 2017). Therefore, the potential of wireless control has not been unlocked yet.

There is a growing trend that jointly considers the control and communication together. The most intuitive reason is that in a wireless network control systems, wireless communication and control are not isolated. Obvious interaction between them will have a crucial effect on the entire system performance. For one thing, the natural characteristics of wireless networks and some specific functions to be implemented can become factors limiting control performance. For another, control systems should also consider the sample period, strict time delay and reliability constraints, which means the message transmissions should be sufficiently reliable and within a specified time limit, and the period should fluctuate in a banded limit (Luvisotto et al., 2017). Traditional communication systems are also indistinguishable from the transmitted data, and the importance is the same, but it is not always the case for control systems. In addition, the total control cost and wireless consumption caused by the separate design of communication and control is another challenge to be considered (Zhao et al., 2019). Therefore, competitive trade-offs among them have to be taken into consideration when designing both the control system and the communication since control system and communication system design are based on different principles (Chamaken and Litz, 2010).

In this chapter, wireless network control (WNC) will be discussed. Background knowledge on industrial automation will be introduced in Section 9.2. Section 9.3 will review the basic concepts of WNC and explains the main points in communication networks and control systems. Section 9.3.2 depicts the fundamental requirements when a WNC model is designed. Section 9.3.3 focuses on the critical variables that have a significant effect on the entire system. Some current practical contributions in terms of communication and control co-design are mentioned in Section 9.4.

9.2 Industrial Automation

The effective deployment of wireless networks at different levels of industrial networks has long been a subject of extensive investigation by the scientific community for several reasons:

1) The direct demands of lowering the installation and maintenance costs that lead to economic savings.
2) Improve the adaptability, resource efficiency as well as the improved integration of supply and demand processes in factories (Gerlach-Erhardt, 2009).
3) Wireless networks can be used for connecting machine parts or machines in difficult or dangerous environments, e.g. over large distances or explosive areas.

Industrial automation is to improve the safety of the production process, production efficiency, product quality, and reduce raw materials and energy consumption in the production process, therefore putting higher requirements on several performance indicators. These indicators can be roughly classified into process automation (PA), building automation (BA), factory automation (FA), power system automation (PSA), power electronic automation (PEC). As Table 9.1 shows, different scenarios have put forward different levels of requirements of real-time performance, data rate and reliability requirements, and proposes some critical indicators in industrial scenarios (Luvisotto et al., 2017).

Due to the market share and cost effects, the most important and mature industrial wireless communication systems in FA mainly use

Table 9.1 Requirements of key performance indicators for different scenarios.

Application Scenarios	Class	Number of nodes	Cycle time per nodes (update rate)	Packet loss rate
Room temperature monitor	Building Automation	1–30	Several seconds	$<10^{-3}$
Close-loop control	Process Automation	1–1000	10–100ms	$<10^{-5}$
Printing machines	Factory Automation	1–100	<2ms	$<10^{-9}$
Substation automation	Power system automation	1–200	50-200us	$<10^{-9}$
Aviation mission electronic system	Power electronic automation	1–50	1–50us	$<10^{-9}$

typically WLAN and Bluetooth standard transceiver components and partly add proprietary protocol extensions. There are some protocols for wireless control network systems, such as WirelessHART, ISA-100.11a, and IEEE 802.15.4e, which are all based on IEEE 802.15.4 standards in wireless personal area networks (WPANs). Others like WIA-FA and IEEE 802.11e are based on the IEEE 802.11 standard in Wireless Local Area Networks (WLAN)s. The problem is that almost all of the existing solutions can provide limited data rates, limited real-time performance, reliability and can only be applied to limited and specific application scenarios, which does not satisfy the ultra-high performance and the new challenges of industry 4.0 (Luvisotto et al., 2017).

9.3 WNC System Model

9.3.1 WNC Model

Figure 9.1 depicts the classical closed loop diagram of a WNC where multiple plants are controlled by a controller through a series of wireless nodes (which can also be used for a relay). It is also called a two-channel feedback network control systems (NCS) since both sensor–controller and controller–actuator channels are applied to this model (Hespanha et al., 2017). It is based on the traditional close

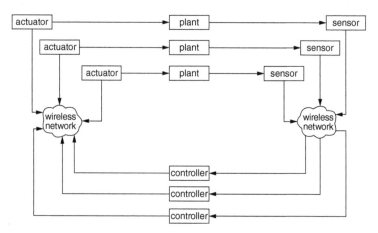

Figure 9.1 A WNC with multiple closed loops.

control system while the plant, actuators and controller are separated. The continuous-time signals associated with the state of the plant are collected from sensors and transmitted to the controller via wireless networks. Once the controller receives the measurements, the control command will be generated and sent to the actuator attached to the plant.

This structure makes it possible to implement a variety of emerging technology models, in particular to provide a potential solution for massive real-time remote control. Robots could be controlled all over the world and healthcare would not be restricted by the location. A telepresence system could be freed with real-time and high reliability haptic feedback, enabling complicated interaction between environments and every participant (Gerlach-Erhardt, 2009).

9.3.1.1 Wireless Networks

The objective of wireless networks is to provide high reliability, high speed, low latency and to work with the control system to ultimately provide satisfactory performance for the entire system. This can be significant different to the widely used network systems like 4G LTE and WIFI. Current solutions for the communication system need to be redesigned to integrate with the control system. In the WNC model, the data transmitted between the sensor and controller tend to be short, with payload sizes of the order of 10^2 bits, as shown in Figure 9.2. In this situation'Cthe Shannon capacity formula in the traditional sense is no longer applicable, and the method of using the coding scheme used by Wi-Fi LTE to reduce the bit error rate is no longer applicable (Durisi et al., 2016). Therefore, the packet structure for the WNC should be redesigned and the channel code method should be rescheduled.

Figure 9.2 A diagram of a short packet and a long packet.

A feasible method obtained by Polyanskiy et al. (2010) showed that the maximum coding rate R* can be expressed as Equation (9.1), which makes the quantitative calculation of the short packet channel capacity possible. It also shows that traditional schemes such as convolutional codes and block codes can provide higher efficiency and shorter decoding delays (Polyanskiy et al., 2010).

$$R^*(n, \varepsilon) = C - \frac{V}{n}Q^{-1}(\varepsilon) + O\left(\frac{\log n}{n}\right).$$ (9.1)

In addition, for channel access, the carrier sense multiple access/collision avoidance (CSMA/CA) mechanism in the distributed coordination function (DCF) of the medium access control (MAC) layer is not suitable since it will cause non-deterministic and long packet delays (Maadani and Motamedi, 2012). The TDMA system appears to be a potential alternative solution, since it allows for bounded channel access delay since a strict timing synchronization and centralized scheduling are used for this scheme to work (Luvisotto et al., 2017). The frequency spectrum should also be reconsidered, and millimetre wave communication is a promising field, since small wavelengths will utilize polarization and spatial processing techniques such as massive MIMO and adaptive beamforming (Ylmaz et al., 2015). Massive multiple-input multiple-output (MIMO) technology can also be considered to apply to WNC since it can ensure high reliability and has already been verified in truly industrial automation environments (Ylmaz et al., 2015).

9.3.1.2 Control System

In addition to meeting the demands of a general control system, such as stability, fast and smooth steady state response, and elimination of steady error, controllers in WNC cope with the challenges caused by the instability of the wireless network system, even actively compensating for the shortage of wireless network to optimize the performance of the entire system. This is mainly achieved by designing a proper controller. A proportional-integral-derivative controller (PID controller) is the most classical control algorithm that could ensure close loop system stable via adjusting the proportion, integer and derivative parameters. The LQR algorithm can be used to design controllers in linear systems (Athans, 1971). The fundamental control system is to consider the linear, time invariant continuous-time system, which is shown by Equations 9.2 and 9.3:

$$\mathbf{x}(t) = \mathbf{A} \cdot \mathbf{x}(t) + \mathbf{B} \cdot \mathbf{u}(t) \tag{9.2}$$

$$\mathbf{y}(t) = \mathbf{C} \cdot \mathbf{x}(t) \tag{9.3}$$

where $x(t) \in \mathbb{R}^n$ is the state vector, $y(t) \in \mathbb{R}^m$ is the output vector and $u(t) \in \mathbb{R}$ is the control input. $\mathbf{A}$ and $\mathbf{B}$ represent the system parameter matrices. The disturbance caused by additive white Gaussian noise (AWGN) with zero mean and variance will always add to these formulas.

The stability and control costs are the two priority indicators that should be considered. *Stability* requires that the control system produces a bounded output for a bound input and ensures a bounded output under a certain degree of disturbance. *Control cost* is used to quantify the close loop performance of a control system by assuming state error zeros and minimizing the control actions (Dorf and Bishop, 2008). The quadratic control cost can also be calculated as a sum of the quadratic functions of the state deviation and the control effort (Athans, 1971).

9.3.2 WNC System Requirements

9.3.2.1 System Structure

For the internal architecture, WNC will have different layers for different functions, including data collection, human interaction and microcontroller control. Different layers may have different protocols or physical media, which may need a gateway to promote communication, and The linear quadratic regulator (LQR) is another algorithm through automotive way to find the optimal linear feedback controller (Athans, 1971; Galloway and Hancke, 2013). There is a trade-off between a flattened network architecture and retaining correlation with the functional hierarchy of the controlled equipment, as shown in Figure 9.3. For communication architecture design the current research mainly focuses on the MAC layer, network layer. The transport function MAC layer is responsible for the establishment, removal and error detection of data links. The network layer is responsible for packet forwarding including routing through intermediate routers The transport layer is the first end-to-end layer

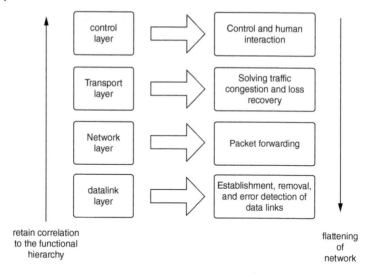

Figure 9.3 The structure of the open system interconnection applied to the WNC.

of the two computers through the network for data communication, by solving congestion and loss recovery; all of them play crucial roles in decreasing latency, increasing reliability, and lowering power consumption. The external architecture refers to the position and distribution of equipment being limited by the power system and electromagnetic conditions.

9.3.2.2 Real-Time Performance

The data need to be transported, dealt with and responded to as soon as possible required by the speed of the device and process (Galloway and Hancke, 2013). However, this performance requirement often contradicts other requirements. A delay would affect the control performance seriously, especially in a closed loop. Real-time performance is the most urgent problem in wireless control systems for some emerging applications. For example, the biggest challenge of tactile internet is to overcome the '1 ms challenge', which requires ultra-low latency. Latency conditions for a haptic response should below 1 ms to avoid delay. Otherwise, the haptic data will not be synchronized with visual and audio signals if the time intervals between visual and tactile movement exceed 1 ms (van den Berg et al., 2017).

9.3.2.3 High Reliability

Wireless control network equipment is often subjected to harsh environmental conditions such as dust, vibration and heat, and equipment must be able to withstand the expected conditions.

Industrial automation will prioritise the transition of safe-critical information as an unexpected message or message delay could cause disastrous consequences. Therefore, these messages should be transmitted in high reliability and low latency (Luvisotto et al., 2017).

9.3.2.4 Determinism

Industrial automation needs messages transmitted in a predictable or deterministic fashion. Low jitter and period fluctuation should be limited due to the fact that variance in time has a negative effect on control loops (Galloway and Hancke, 2013). The problem will be even worse in the case of the exchange of data between a central controller and a set of distributed sensors and actuators

9.3.2.5 Sample Data Traffic and Event Order

Industrial networks should enable transmitting both period sample data and aperiodic events and thus priority has a significant effect on the latency. The sampling period depends on the types of sensors and will be interrupt by sudden events. Figure 9.4 shows a circumstance where the path of the node needs to be planned in a multi-hop wireless NCS (WNCS). One scheduled example is to allow the transmission of a node to be prescheduled beforehand to plan in advance which

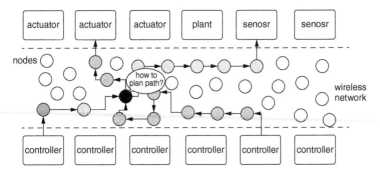

Figure 9.4 The circumstance where the path of the node needs to be planned in a multi-hop WNCS.

nodes of the data transmission path should avoid traffic congestion and disconnection in the communication path (Sadi and Ergen, 2013).

9.3.3 Analysis of Influencing Factors

9.3.3.1 Sampling Period

WNCS requires time triggered sampling for periodic data and event-triggered sampling for aperiodic events. Based on the communication and control co-design idea, a higher sample rate results in better control performance but on the other hand high sample rates tend to increase the network traffic in wireless networks since increasing network load in turn increases the message loss probability and message delay and finally degrades the control performance (Heemels et al., 2012). 'Best effort' for the existing commercial networks is not suitable for the WNC system.

9.3.3.2 Time Delay

Time delays mainly occur in the sensor–controller and controller–actuator. The random delay occurred in commercial networks is the trickiest problem to be solved in WNCs. In terms of the final control performance the increase in both delays will degrade system performance, which will finally cause phase shifts that limit the control bandwidth and affect closed-loop stability (Wittenmark et al., 2002). Delay mainly occurs in data transport, medium access and queues in communication systems. Computing time cost, the time spent on analogue to digital and digital to analogue will also cause time delays. The total time delay should not be an overhead to the sample period, or the packet will be considered to be lost. Figure 9.5 illustrates the potential time delay that would happen

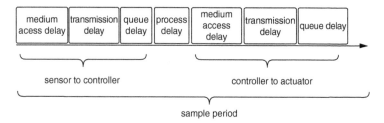

Figure 9.5 Potential time delay diagram in the WNCS.

in the WNCS. In addition, time-varying channels, a limited spectrum, and all kinds of interference, signal attenuation, multi-path effects and packet retransmissions, will make this situation worse (Chamaken and Litz, 2010).

9.3.3.3 Packet Loss

Packets may be lost during transmission due to both internal and external effects.

Multi-path, doppler shift and interference have a major effect on the communication system that will degrade the control system performance and even cause disastrous consequences (Park et al., 2018).

9.4 Network and System Control Co-design

Significant contributions have been made both in the fields of communication and control. Based on the quality of service (QoS) of the communication system, the corresponding control system is designed. In the meantime, the communication system could also be designed based on the QoS of the control system. However, the problem is whether to focus on one aspect or optimize the other, which is not in the true sense of joint design. Chamaken and Litz (2010) give a feasible co–design idea that leaves room for design communication and control, and then optimize the important variables according to the specific QoS and required degrees of freedom. This reduces the difficulty of joint design to a certain extent but is based on a certain degree of compromise, so it does not fundamentally solve the problem.

The current research on co-design mainly focuses on the above mentioned critical variables by studying the interaction between several variables and designing adaptive algorithms to find the best system effect. Early research has been made in analysing the interactive effects of network induced delay and packet loss on the control system performance (Gaid et al., 2006; Safari and Tavassoli, 2014). Shan et al. (2011) proved that the necessary trade-off is needed to consider among data rate, packet loss and time delay are intricate and implicit in the control performance to achieve final optimization and a cross-layer design framework and corresponding protocols are proposed to achieve optimizing performance.

Some practical communications and control co-designs are based on the optimization some of the critical variables. Colandairaj et al. (2007) focused on achieving the optimization of the data rate scaling in conjunction with sample rate adaption and the control algorithm to adapt the sampling interval. The reduction in channel bandwidth is compensated by a reduction in network traffic. A novel time-based finite horizon optimal control and communication co-design was proposed by adopting the emerging neuro-dynamics programming (NDP) technique and actor–critic–identifier (ACI) architecture in (Xu et al., 2016). The dynamics of NCS was reduced and the effectiveness of power constraint was satisfied by applying the Lyapunov Theorem. Xu and Carrillo (2017) focused on resisting the disturbance and reducing the effects caused by network delay by designing an augmented system; a synthesis controller was redesigned. Packetised predictive network control (PPC) is another promising method for optimizing the sequence of control inputs sent when data loss happens in a communication network (Xu and Carrillo (2017), which is mentioned in Chapter 1.

9.5 Conclusion

In this chapter, we have discussed the fundamental design capabilities needed to realize future wireless network control. WNCs are increasingly being used in many emerging applications, and will play a crucial role in industrial automation in the near future. It is worth noting that because of the close integration of communication and control systems, the idea of joint design is gradually being implemented, and some ideas in WNCs are about to become a reality.

Bibliography

Ahlgren B., Hidell M. and Ngai E. C. (2016) Internet of Things for Smart Cities: Interoperability and Open Data. *IEEE Internet Computing*, 20, 6, 52–56.

Athans M. (1971) The role and use of the stochastic linear-quadratic-gaussian problem in control system design. *IEEE Transactions on Automatic Control*, 16, 6, 529–552.

Chamaken A. and Litz L. (2010) Joint design of control and communication in wireless networked control systems: A case study. *Proceedings of the 2010 American Control Conference, Baltimore, MD, 2010*, 1835–1840.

Colandairaj J., Irwin G. W. and Scanlon W. G. (2007) A Co-Design Solution for Wireless Feedback Control. *2007 IEEE International Conference on Networking, Sensing and Control, London, 2007*, 404–409.

Dorf R. C. and Bishop R. H. (2011) *Modern Control Systems*, Pearson Education, London.

Durisi G., Koch T. and Popovski P. (2016) Toward massive, ultrareliable, and low-latency wireless communication with short packets. *Proc. IEEE*, 104, 9, 1711–1726.

Farhangi H. (2010) The path of the smart grid. *IEEE Power and Energy Magazine*, 8, 1.

Gaid M. E. M. B., Cela A. and Hamam Y. (2006) Optimal integrated control and scheduling of networked control systems with communication constraints: application to a car suspension system. *IEEE Transactions on Control Systems Technology*, 14, 4, 776–787.

Galloway B. and Hancke G. P. (2013) Introduction to Industrial Control Networks. *IEEE Communications Surveys & Tutorials*, 15, 2, 860–880.

Gerlach-Erhardt H. (2009) Real time requirements in industrial automation. PNO TC2/WG12, ETSI Wireless Factory Starter Group Meeting, SpringerOpen.

Heemels W. P. M. H., Johansson K. H. and Tabuada P. (2012) An introduction to event-triggered and self-triggered control. *IEEE CDC*.

Hespanha J. P., Naghshtabrizi P. and Xu Y. (2007) A survey of recent results in networked control systems. *Proceedings of the IEEE*, 95, 1, 138–162.

Lee E. A. (2008) Cyber Physical Systems: Design Challenges. *2008 11th IEEE International Symposium on Object and Component-Oriented Real-Time Distributed Computing (ISORC), Orlando, FL, 2008*, 363–369.

Luvisotto M., Pang Z. and Dzung D. (2017) Ultra High Performance Wireless Control for Critical Applications: Challenges and Directions. *IEEE Transactions on Industrial Informatics*, 13, 3, 1448–1459.

Maadani M. and Motamedi S. A. (2012) An adaptive retry-limit scheme for real-time IEEE 802.11 based wireless industrial networks. *7th International Conference on Computer Science & Education (ICCSE)*, Melbourne, VIC, 2012, 267-271.

Park P. , Coleri Ergen S., Fischione C., Lu C. and Johansson K. H. (2018) Wireless Network Design for Control Systems: A Survey. *IEEE Communications Surveys & Tutorials*, 20, 2, 978–1013.

Polyanskiy Y., Poor H. V. and Verdu S. (2010) Channel coding rate in the finite blocklength regime. *IEEE Transactions on Information Theory*, 56, 5, 2307–2359.

Quevedo D. E., Silva E. I. and Goodwin G. C. (2007) Packetized Predictive Control over Erasure Channels. *2007 American Control Conference, New York, NY, 2007*, 1003–1008.

Sadi Y. and Ergen S. C. (2013) Optimal power control, rate adaptation, and scheduling for UWB-based intravehicular wireless sensor networks. *IEEE Transactions on Vehicular Technology*, 62, 1, 219– 234.

Safari R. and Tavassoli B. (2014) Stability analysis of networked control systems with generalized nonlinear perturbations *2014 Smart Grid Conference (SGC), Tehran, 2014*, 1–6.

Shan H., Cheng H. T. and Zhuang W. (2011) Cross-Layer Cooperative MAC Protocol in Distributed Wireless Networks. *IEEE Transactions on Wireless Communications*, 10, 8, 2603–2615.

Simsek M., Aijaz A., Dohler M., Sachs J. and Fettweis G. (2016) 5G-enabled tactile internet. *IEEE J. Selected Area Comm.* 34, 3, 460–473.

van den Berg D., Glans R., De Koning D. et al. (2017) Challenges in Haptic Communications over the Tactile Internet. IEEE Access, 5, 23502–23518.

van den Berg D. et al (2017) Challenges in haptic communications over the Tactile Internet. *IEEE Access*, 5, 23502–23518.

Wittenmark B., Astro m K. J. and Arzen K.-E. (2002) Computer control: An overview. *IFAC Professional Brief, Technical Report*.

Xu H. (2016) Finite horizon optimal control and communication co-design for uncertain networked control system with transmit power constraint. *2016 IEEE Symposium Series on Computational Intelligence (SSCI), Athens, 2016*, 1–7.

Xu H. and Carrillo L. R. G. (2017) Near optimal control and network co-design for uncertain networked control system with constraints. *2017 American Control Conference (ACC), Seattle, WA, 2017*, 2339–2344.

Yilmaz O. N. C. et al. (2015) Analysis of ultra-reliable and low-latency 5G communication for a factory automation use case. *Proc. 2015 EEE International Conference on Communications, June 2015*, 1190–1195.

Zhao G., Imran M. A., Pang Z., Chen Z. and Li L. (2019) Toward Real-Time Control in Future Wireless Networks: Communication-Control Co-Design. *IEEE Communications Magazine*, 57, 2, 138–144.

10

Caching at the Edge in Low Latency Wireless Networks

Ramy Amer[1], M. Majid Butt[2], and Nicola Marchetti[3]

[1] CONNECT Centre for Future Networks, Trinity College Dublin, Ireland
[2] Nokia Bell Labs, France, and CONNECT Centre for Future Networks, Trinity College Dublin, Ireland
[3] CONNECT Centre for Future Networks, Trinity College Dublin, Ireland

10.1 Introduction

The recent proliferation of smartphones has substantially enriched the mobile user experience, leading to a vast range of new wireless services, including video-on-demand (VoD) streaming, web browsing, and social networking. This phenomenon has been further dominated by mobile video streaming, which currently accounts for almost 50 % of mobile data traffic with an estimate of a 500-fold increase over the next few years (Cisco, 2016). Additionally, the number of connected devices and data rates are anticipated to grow exponentially, due to several new applications and technologies beyond personal communications (Andrews et al., 2014), e.g., vehicular networks (Baza et al., 2019; Baza, 2019) and smart grids (Baza et al., 2018). These new challenges have encouraged mobile network operators (MNOs) to redesign their current networks into more advanced and sophisticated ones that can increase coverage, reduce latency, boost capacity, and bring contents near to the users in a cost effective way.

One of the promising approaches to meet such unprecedented traffic demands is the deployment of SCNs (Andrews, 2013). Small cell networking is a novel paradigm based on the idea of deploying short-range small base stations (SBSs) underlaying the existing macrocellular network. Regarding SCNs, to date, the vast majority of research has dealt with issues related to self-organization, inter-cell interference coordination (ICIC), traffic offloading, and energy efficiency (EE), see (Andrews, 2013) and the references therein. These studies are carried out under the existing reactive networking paradigm wherein users' traffic requests are served upon their arrival or dropped in case of outages. Owing to these characteristics of the reactive network, the existing SCN paradigm does not help to accommodate the peak traffic demands from the network. This shortcoming is set to become more severe, in particular because of the surging number of connected devices and the advent of ultra-dense networks (UDNs), which will continue to exhaust current cellular network infrastructures. Different from the evolutionary path of previous cellular communication generations that was based on spectral efficiency improvements, the most substantial system performance gains will be obtained by means of network infrastructure densification. By increasing the density of operator-deployed infrastructure elements, along with incorporation of user-deployed access nodes and mobile user devices acting as 'infrastructure prosumers', i.e. when users' devices are employed in communication as relays, it is expected that having one or more access nodes exclusively dedicated to each user will become feasible. Although it is clear that UDNs are able to take advantage of the significant benefits provided by proximal transmissions and increased spatial reuse of system resources, at the same time large node density and irregular deployments introduce new challenges. These challenges are mainly due to interference characteristics. In addition, the backhauling of dense small cell networks has emerged as a bottleneck for their successful deployment. The increasing number of deployed small cells and the lack of high capacity backhaul links, still represent limiting factors to the network densification gains. Therefore, it becomes of paramount importance to envisage a new networking paradigm that goes beyond current heterogeneous small cell deployments, i.e. introducing networks with different types of SBSs with different transmission powers and capabilities, leveraging the latest developments in storage, context awareness, and social networking (Trial, 2012).

Cellular networks, increasingly the most essential aspect of our telecommunication infrastructure, are in a period of unprecedented change. Hence, incremental changes in designing and optimizing such reactive networks are becoming more and more outdated. Future cellular networks are expected to be smart in the sense that network nodes anticipate users' demands and utilize predictive abilities to reduce the traffic peak-to-average ratio, which yields significant network resource savings. Meanwhile, proactive caching in the wired internet is well established and has been shown to reduce latency and energy consumption (Liu et al., 2016). Similar benefits can be expected by caching popular contents at the wireless edge, which can improve the network performance and accommodate the explosive demand for wireless data. Proactive caching either at users with device-to-device (D2D) communications enabled, or at SBSs to eliminate the backhaul bottleneck, is envisioned as a promising solution to satisfy the high demand for data and to alleviate the heavy burden on the core network. This is particularly useful because some of the content is available locally, instead of requiring redundant traverses across backhaul links (Bastug et al., 2014; Golrezaei et al., 2013; Ji et al., 2016a; Liu and Yang, 2016; Maddah-Ali and Niesen, 2014).

Starting from describing the main features of wireless caching networks, this chapter extends the discussion to the proposed inter-cluster cooperative caching architecture for wireless D2D networks (Amer et al., 2018a) aiming at minimizing the average service delay per content request. This chapter first introduces an overview of the literature on wireless caching networks and states the main challenges. Then, our proposed caching architecture and algorithms are introduced to enhance the performance of wireless caching networks and address these challenges.

This chapter is organized in the following way. Section 10.2 describes the main characteristics of wireless caching networks. Section 10.3 then discusses different architectures, approaches, and content placement schemes for wireless caching networks. Section 10.4 states the role of wireless caching in low-latency wireless networks, and Section 10.5 introduces our proposed caching architecture for D2D clustered networks. Section 10.6 presents numerical results and performance evaluation, and Section 10.7 concludes the chapter.

10.2 Living on the Edge

The wireless caching network, as described in the seminal work (Bastug et al., 2014), is proactive in essence and rooted in the fact that network nodes, including base stations (BSs), and mobile devices, would exploit users' context information, foresee users' demands, and leverage their predictive abilities to achieve significant resource savings, sustain acceptable quality of service (QoS) levels and keep low cost/energy expenditure (Bastug et al., 2013). This paradigm goes far beyond current cellular deployments, which have been designed mainly to serve 'dumb' devices with very limited storage and processing power. In fact, current smartphones have become very sophisticated with considerably improved computing and storage capabilities. Subsequently, under the proactive networking paradigm, network nodes are assumed to track, learn, and build users' demand profiles to predict future requests, leveraging devices' capabilities and the vast amount of available data.

Recently, predictive analytics have received significant attention using machine learning techniques to analyse billions of infrastructure logs to produce predictive and actionable information for outage prediction and content recommendation (Etter et al., 2012). Leveraging these predictive capabilities, users can be scheduled in a more efficient manner, and resources are pre-allocated more intelligently by proactively serving peak-hour demands during off-peak times (e.g., at night). The proactive caching paradigm leverages the statistical traffic patterns and users' context information (i.e. file popularity distributions, location, velocity, and mobility patterns) aiming at a better prediction of when users' content is requested, and matching that with the amount of resources needed, and at which network locations specific content should be stored (cached). As a relevant practical example, online social networks (e.g., Facebook, Twitter) have become instrumental in users' content distribution (Robinson et al., 2014). Users tend to value highly recommended content by friends or people with similar interests and are also likely to recommend them. Figure 10.1 shows an example of a spatial network layer overlaid with the social network layer.

Now we are in a position to discuss different architectures for wireless caching networks. Caches can be installed in a macro base station (MBS), SBS (say, pico or femto BS), relay nodes, and user devices (Liu et al., 2016), see Figure 10.2. A picocell is a small cellular BS typically covering a small area, such as in-building (offices, shopping malls, train stations, stock exchanges, etc.), while a femtocell is defined as

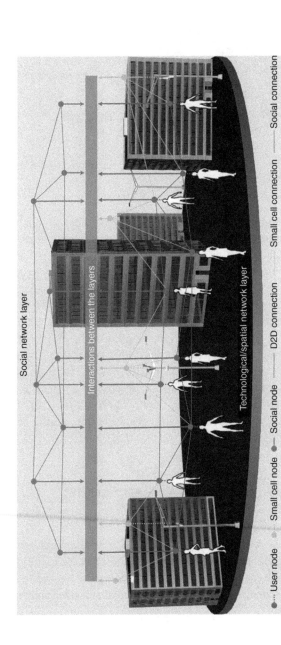

Figure 10.1 An illustration of an overlay of a socially interconnected and technological/spatial network (Bastug et al., 2014).

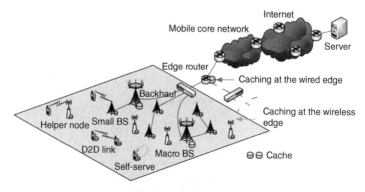

Figure 10.2 An illustration of local caching and content delivery at the wireless edge.

a small, low-power cellular BS, typically designed for use in a home or small business. Compared to caching at the evolved packet core (EPC), caching at existing MBSs and SBSs essentially plays the role of replacing backhaul links, and hence alleviates backhaul congestion. Moreover, a new type of SBS with weak backhaul connections, called helpers (relays) (Golrezaei et al., 2013), can enable flexible and cost-effective deployments to deliver popular content. Due to the fact that increasing the cache size can increase the cache-hit probability, and hence lower the required backhaul capacity, there is a trade-off between cache size and backhaul capacity. Besides, caching content at user terminals such as smartphones, tablets, and laptops has been applied as a technique to improve quality of experience (QoE) (Higgins et al., 2012), and recently, it has also been proposed to offload wireless traffic.

10.3 Classifications of Wireless Caching Networks

This section is devoted to the discussion of three main architectures for wireless caching networks, namely, caching on SCNs, caching over the C-RAN, and caching among the mobile devices. Then, we describe two possible schemes of content placement that are widely adopted in the literature.

10.3.1 Wireless Caching Architecture

We here classify the wireless caching network from the architecture perspective.

- Small cell networks: The introduction of small cell BSs is viewed as a key paradigm to handling the increase in video traffic and improve wireless capacity by bringing content closer to users. However, reaping the benefits of small cell deployment requires meeting several key challenges such as resource allocation and network modelling. For instance, owing to cheap storage/memory prices and the fact that mobile video accounts for most of the internet traffic demands, one can leverage the use of storage at the small cell level to bring popular content closer to the network edge (i.e. BSs) (Baştu et al., 2015). Indeed, one promising approach to improve the QoS of video transmission is through caching popular contents locally at the SBSs to alleviate peak traffic demands and minimize service delays. (Shanmugam et al., 2013).

- Cloud radio access networks: While content delivery networks (CDNs) have been recently enhanced to reduce internet bandwidth consumption and associated delay/jitter of online video streaming, such content consumed by mobile devices must additionally travel through the wireless carrier core network and radio access network (RAN) before reaching the user equipment (UE) (Ahlehagh and Dey, 2012). One promising approach to address the need for massive content distribution is to enable content caching within the wireless operators' networks, where popular contents are cached at the BS at the edge of the RAN (Tran et al., 2017). C-RAN is an emerging architecture 5G wireless system, in which a centralized baseband unit (BBU) implements the baseband processing functionalities for a set of RRHs, which are connected to the BBU by means of fronthaul links (Checko et al., 2015; Simeone et al., 2016). Recently, an evolved network architecture, referred to as a fog radio access network (F-RAN), was proposed, which enhances the C-RAN architecture by allowing the RRH to be equipped with caching and signal processing functionalities (Bi et al., 2015; Peng et al., 2016; Pfeiffer, 2015). This architecture is also referred to as a hybrid of cloud and fog processing in the literature (Chiang and Zhang, 2016). As a cache-aided system, an F-RAN operates in two phases, namely the pre-fetching and the delivery phases (Maddah-Ali and Niesen, 2015; Tao et al., 2016). Pre-fetching operates at the large time scale corresponding to the period in which content popularity remains constant. This time scale encompasses multiple transmission intervals. Based on the cached file messages, the delivery phase, instead, operates separately on each transmission interval. The fronthaul-aware design of the pre-fetching or delivery phases was studied in (Azari et al., 2016; Peng et al., 2015; Tandon and Simeone, 2016; Tao et al., 2016) under the assumption that the fronthaul links in an F-RAN are leveraged to transport requested content (not present in the local caches) to the RRHs.

- Device-to-device communication: Capitalizing on the fact that user demands are highly redundant, each user demand can be satisfied through local communication from the device's cache, without requiring a high throughput backhaul to the core network (Ji et al., 2016b). The concept of having helper nodes is pushed further by introducing the notion of wireless devices as helpers (Golrezaei et al., 2013). Recent years have seen an enormous proliferation of smartphones and tablets that have anywhere between 10 to 64 GB of storage (not to mention the 500 GB on typical laptop hard disks). By enabling D2D communications, the ensemble of wireless devices can become a distributed cache that allows a more efficient downloading of content as compared to traditional networks without caching. The advantage of using wireless devices instead of fixed helper nodes lies in the small deployment costs and automatic upscaling of the capacity as the density of such devices increases. The drawback lies in the necessity to motivate users to participate in the cache, and the randomness of the available throughput due to the decentralized and uncoordinated nature of D2D communication. More relevantly, some works focus on the economic aspect of caching in wireless D2D networks. In such networks, the operators define a pricing scheme to motivate users to proactively download the most popular files and cache them in their devices to serve other users' requests. In (Hosny et al., 2015), the authors proposed a smart pricing scheme to maximize the benefit of the operator and minimize the charged price to the users. Via D2D communications, users can trade their cached files to minimize their expected payments. On the other hand, the operator defines a dynamic pricing model that differentiates off-peak and peak time periods to maximize its own benefit. If a user requests one of the files stored in neighbours' caches in the cluster, the neighbours will handle the request locally through D2D communication (Amer et al., 2018c, 2019). Otherwise, the BS should serve the request. As a result, the probability of having more D2D communications among the users depends on what is stored by users (Golrezaei et al., 2013).

Prior works in the literature show that the file placement largely follows two approaches: deterministic placement and probabilistic placement. For deterministic placement, files are cached and optimized for specific networks in a deterministic manner (Amer et al., 2018a; Golrezaei et al., 2014b). However, in practice, the wireless channels and the geographic distribution of devices are time variant. This triggers the optimal file placement strategy to be frequently updated, which makes the file placement highly complicated. To cope with this problem, probabilistic file placement considers that each device randomly caches a subset of content with a certain caching probability (Amer et al., 2018b, 2019; Blaszczyszyn and Giovanidis, 2015).

10.4   Caching For Low Latency Wireless Networks

Edge processing is one of the emerging trends in the evolution of 5G networks (Sengupta et al., 2016). It refers to utilization of locally stored content and computing resources at the network edge, i.e. closer to the users. Such localization is particularly appealing for both low-latency or location-based applications as well as multimedia transmissions. In particular, pushing caching and computing resources to the network edge has been shown to significantly reduce latency (Bennis et al., 2018). This is attributed to the fact that local availability of popular content at the network edge has the potential of reducing the delivery latency as well as the overhead on backhaul connections to content servers. As a result, the role of cache-enabled networks in reducing service latency has been studied extensively in recent literature (Golrezaei et al., 2012; Ji et al., 2015; Maddah-Ali and Niesen, 2014; Sengupta et al., 2015).

The caching capacity of local BSs can be regarded as a new type of resource available to wireless networks besides time, frequency, and space (Peng et al., 2015). However, it is limited when compared to the total amount of mobile traffic. Thus sophisticated caching placement strategies are needed to fully exploit the benefit of caching. With the knowledge of channel statistics and file popularity distributions, a central controller of a wireless network is able to determine an optimal caching strategy to cater for user requests with locally cached content. Once caches are fully utilized, the requirements for backhaul can be greatly reduced and the download delay can be shortened, especially when the backhaul links are in poor condition.

We have already discussed the emergence of edge caching as a promising approach to alleviate the heavy burden on data transmission and allow for reliable low latency communication. However, deploying caches in wireless networks poses many new challenges. Motivated by the above discussion, in the rest of this chapter we present a new caching architecture, which is shown to improve the network performance in terms of the average service delay. In particular, we envision a new D2D caching architecture by allowing D2D communication alongside inter-cluster cooperation.

10.5 Inter-cluster Cooperation for Wireless D2D Caching Networks

The architecture of device caching exploits the large storage available in modern smartphones to cache multimedia files that might frequently be requested by the users. The users' devices exchange multimedia content stored on their local storage with nearby devices (Ji et al., 2016b). Since the distance between the requesting user and the caching user (a user who stores the file) will be small in most cases, D2D communication is commonly used for content transmission (Ji et al., 2016b). For instance, (Golrezaei et al., 2014a) proposed a novel architecture to improve the throughput of video transmission in cellular networks based on the caching of popular video files in BS controlled D2D communication. The analysis of this network is based on the subdivision of a macrocell into small virtual clusters such that one D2D link can be active within each cluster. Random caching is considered where each user caches files at random and independently, according to certain caching distribution.

Motivated by the remarks from the above discussion, i.e. backhaul links being highly congested, the geometric distribution of the users as groups in clusters, and the small memory sizes of a group of users collocated in the same cluster, we propose a novel D2D caching architecture with inter-cluster cooperation. We propose a system in which a user in a given cluster can search for its requested files either in the local cluster or any of the remote clusters. We show that allowing inter-cluster collaboration via cellular communication achieves both user and system performance gains. From the user perspective, the average service delay per-request is reduced when downloading files from a remote cluster instead of getting files from the core network. From the system perspective, the heavy burden on backhaul links is alleviated by decreasing the number of requests that are served directly from the core network. We analyse the network average delay and outage probability for the proposed inter-cluster cooperative caching system under different caching schemes and show how the network performance is significantly improved.

10.5.1 Proposed Network Model

Next, we describe our proposed D2D caching network with inter-cluster cooperation. Fig. 10.3 illustrates the system layout. A

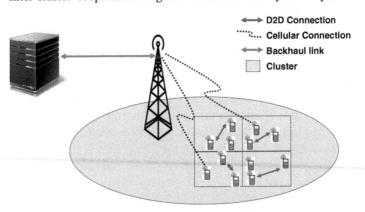

Figure 10.3 Schematic diagram of the proposed system model. A cell is divided into square clusters, where devices can download their requested files using D2D, cellular, or backhaul communication.

cellular network consists of an SBS and a set of devices $\mathcal{V} = \{1, \dots, n\}$ placed uniformly in the cell. The cell is divided into a set of equally sized clusters $\mathcal{K} = \{1, \dots, K\}$. For mathematical convenience, we assume that the number of devices per cluster is $y = n/K$ devices, as in (Chen et al., 2017) and references therein. Devices in the same cluster can communicate directly using low-power high-rate D2D communication in a dedicated frequency band for D2D transmission.

Each user $u \in \mathcal{V}$ requests a file f from a file library $\mathcal{F} = \{1, \dots, m\}$ independently and identically, according to a given request probability mass function. It is assumed that each user can cache up to M files, and for the caching problem to be non-trivial, it is assumed that $M < m$. From the cluster perspective, there exists a cluster virtual cache center (VCC) formed by the union of the devices' storage in the same cluster, which caches up to N files, i.e. $N = (n/K)M$.

We assume that the D2D communication does not interfere with the communication between the BS and devices. We also assume that all D2D links share the same time-frequency transmission resource within one cell. Multiple transmissions on those resources are possible since the distance between requesting devices and devices with the stored file will typically be small. Furthermore, there should be no interference by other transmissions on an active D2D link. To achieve this, the cell is divided into smaller areas, which we denote as clusters. To avoid intra-cluster interference, only one such communication per cluster is allowed. Devices in the same cluster are assumed to be served in a round-robin manner.

We define three modes of operation according to how a request for content $f \in \mathcal{F}$ is served:

1) Local cluster mode (M_{lc} mode): Requests are served from the local cluster. Files are downloaded from nearby devices via a single-hop D2D communication. In this mode, we neglect self-caching, i.e. the event when a user finds the requested file in its internal cache with zero delay. Within each cluster, the BS can help devices find their requested content by broadcasting signals containing the content replication ratio.

2) Remote cluster mode (M_{rc} mode): Requests are served from any of the remote clusters via inter-cluster cooperation. The BS fetches the requested content from a remote cluster, then delivers it to the requesting user by acting as a relay in a two-hop cellular

transmission. The BS assists in content dissemination in the remote cluster mode by relaying the content between different clusters.

3) Backhaul mode (M_{bh} mode): Requests are served directly from the backhaul. The BS obtains the requested file from the core network via the backhaul link and then transmits it to the requesting user.

In each cluster, we assume that the stream of user requests are served sequentially based on a first in first out (FIFO) criterion. The BS receives all requests and works as a coordinator to establish the file transfer between the requesting user (a user who requests the file) and the serving node (another user who caches the file or a caching server in the core network). The BS keeps track of which devices can communicate with each other and which files are cached on each device. Such BS-controlled D2D communication is more efficient and more acceptable to spectrum owners if the communication occurs in a licensed band as compared to traditional uncoordinated peer-to-peer communications (Caire and Molisch, 2015). To serve a request for file f in cluster $k \in \mathcal{K}$, first the BS searches the VCC of cluster k. If the file is cached, it will be delivered from the local VCC (M_{lc} mode). We assume that the BS has all the information about cached content in all clusters, such that all file requests are sent to the BS, then the BS replies with the address of the caching user from whom the file will be retrieved.

If a file is not cached locally in cluster k and is instead cached in any of the remote clusters, it will be fetched from a randomly chosen cooperative cluster (M_{rc} mode), rather than downloading it from the backhaul. Unlike multi-hop D2D cooperative caching discussed in (Jeon et al., 2017), cooperating clusters are assumed to exchange cached files using a two-hop cellular communication link through the BS, such that the D2D band is dedicated only to intra-cluster communication. Hence, all the inter-cluster communication is performed in a centralized manner through the BS. Finally, if the requested file has not been cached in any cluster $j \in \mathcal{K}$ in the cell, it can be downloaded from the core network via the backhaul link (M_{bh} mode). The selection of the three modes of operation is conducted in a prioritized order from the local cluster, from the remote cluster, or finally from the core network through the backhaul link as a last resort.

Serving files sequentially according to the above three modes is based on the assumption that the BS has a capacity limited wired backhauling, such that the average delay per request is decreased when allowing inter-cluster cooperation. Otherwise, if the backhaul is not a bottleneck, e.g., optical fibre or millimetre wave backhaul links are available, requests for files not cached in the local cluster are served directly from the core network through the high capacity backhaul link. The analysis in this chapter relies on a well known grid-based clustering model (Golrezaei et al., 2014a), i.e. no specific underlying physical model or parameters are assumed. Therefore, the obtained design/results, e.g., design of caching scheme and the performance of the greedy algorithm, can be applied to similar scenarios with three prioritized paths (modes) for file downloading. For example, on-board devices, such as on a plane or a ship, can obtain requested files from neighbouring devices via Bluetooth (local cluster mode), from a remote user through an access point (AP) acting as a relay (remote cluster mode), or finally from the backhaul, which is the least preferred option.

10.5.2 Content Placement and Traffic Characteristics

We use a binary matrix $\mathbf{C} = [c_{k,f}]_{K \times m}$ with $c_{k,f} \in \{0, 1\}$ to denote the cache placement in all clusters, where $c_{k,f} = 1$ indicates that content f is cached in cluster k. Figure 10.4 shows the assumed devices' traffic

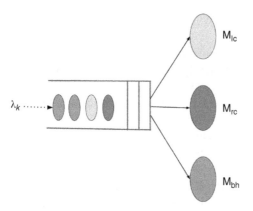

Figure 10.4 The devices' traffic model in a cluster k with cache center VCC is modelled as a multi-class processor sharing queue.

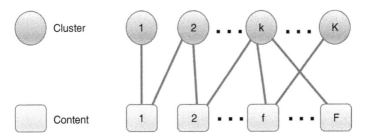

Figure 10.5 An example of the content cache placement modelled as a bipartite graph indicating how files are cached in clusters.

model in a cluster k, modelled as a multi-class processor sharing queue (MPSQ) with arrival rate λ_k, and three serving processors representing the three transmission modes. According to the MPSQ definition (Karray and Jovanovic, 2013), each transmission mode is represented by an M×M×1 queue with Poisson arrival rate and exponential service rate. A graphical interpretation of the content cache placement is shown in Figure 10.5. The content caching policy is defined by a bipartite graph $\mathcal{Y} = (\mathcal{K}, \mathcal{F}, \mathcal{E})$, where edges $(k, f) \in \mathcal{E}$ denote that content f is cached in the VCC of cluster k.

If a user in cluster k requests a locally cached file f (i.e. $c_{k,f} = 1$), it will be served by the local cluster mode with an average rate R_D. However, if the requested file is not cached locally and cached in any of the remote clusters, i.e. when $c_{k,f} = 0$ and $\sum_{j \in \mathcal{K} \setminus \{k\}} c_{j,f} \geq 1$, it will be served by the remote cluster mode.

We denote the rate for the remote cluster mode by R_{WL}, accounting for the average sum transmission rate between the clusters cooperating through the BS. Accordingly, R_{WL} is shared between clusters simultaneously served by the remote cluster mode. Finally, requests for files that are not cached in the entire cell, i.e. when $\sum_{j=1}^{K} c_{k,f} = 0$, are served via the backhaul mode with an average sum rate R_{BH}. We assume that $R_{BH} \ll R_{WL}$, such that the part of the cellular rate allocated to the devices served by the backhaul mode is neglected for the delay analysis. R_{BH} is assumed to be the effective rate from the core network to the end user.

Due to traffic congestion in the core network and the transmission delay between cooperating clusters, we assume that the aggregate transmission rates for the above three modes are ordered such that $R_D > R_{WL} > R_{BH}$. We also assume that the content size S_f is exponentially distributed with mean $\bar{S}$ bits. Hence, the corresponding request service times of the three transmission modes also follow an exponential distribution with means $\tau_{lc} = \frac{\bar{S}}{R_D}$ s, $\tau_{rc} = \frac{\bar{S}}{R_{WL}}$ s, and $\tau_{bh} = \frac{\bar{S}}{R_{BH}}$ s, respectively.

10.5.3 Caching Problem Formulation

In this subsection, we characterize the network average delay on a per request basis from the global network perspective. We assume that the popularity distribution of files in all clusters follows a Zipf distribution with skewness order β (Breslau et al., 1999). However, it is assumed that the content may vary across clusters. This is inspired by the fact that, for instance, devices in a library may be interested in an entirely different set of files from the devices in a sports centre. Our assumption for the popularity distribution is extended from (Ji et al., 2016a), where the authors explained that the scaling of popular files is sublinear with the number of devices.

The probability that a file f is requested in cluster k, with m_0 highly demanded files in each cluster, follows a Zipf distribution written as,

$$P_{k \cdot f} = \frac{(f - \frac{k-1}{k}m_0 a + (m - \frac{k-1}{k}m_0)b)^{-\beta}}{\sum_{i=1}^{m} i^{-\beta}}, \tag{10.1}$$

where $a = \mathbb{1}(f > \frac{k-1}{k}m_0)$ and $b = \mathbb{1}(f \le \frac{k-1}{k}m_0)$, $\frac{k-1}{k}m_0$ is the order of the most popular file in the kth cluster, and $\mathbb{1}(.)$ is the indicator function. When $k = 1$, we get $P_{1 \cdot f} = \frac{(f)^{-\beta}}{\sum_{i=1}^{m} i^{-\beta}}$ for the first cluster, which is the Zipf distribution with the most popular file $f = 1$. For example, if $m_0 = 60$, then $P_{2 \cdot f} = \frac{(f - 30a + (m-30)b)^{-\beta}}{\sum_{i=1}^{m} i^{-\beta}}$ for the second cluster, which is the Zipf distribution with the most popular file $f = \frac{m_0}{2} + 1 = 31$; then $f = \frac{2m_0}{3} + 1 = 41$ is the most popular file in the third cluster, and so on.

10.5.3.1 Arrival and Service Rates

The arrival rates for the three communication modes M_{lc}, M_{rc}, and M_{bh} in a cluster k are denoted respectively by $\lambda_{k,lc}$, $\lambda_{k,rc}$, and $\lambda_{k,bh}$ while the corresponding service rates are represented by μ_{lc}, μ_{rc}, and μ_{bh}. For the local cluster mode, we have

$$\lambda_{k \cdot lc} = \lambda_k \sum_{f=1}^{m} P_{k \cdot f} c_{k \cdot f}, \tag{10.2}$$

where $\sum_{f=1}^{m} P_{k,f} c_{k,f}$ is the probability that the requested file is cached locally in cluster k. The corresponding service rate is $\mu_{lc} = \frac{1}{\tau_{lc}}$. For the remote cluster mode, the request arrival rate is defined as

$$\lambda_{k,rc} = \lambda_k \sum_{f=1}^{m} P_{k,f} (1 - c_{k,f}) \min\left(\sum_{j \in \mathcal{K}\backslash\{k\}} c_{j,f}, 1 \right), \tag{10.3}$$

where $\min(\sum_{j \in \mathcal{K}\backslash\{k\}} c_{j,f}, 1)$ equals one only if the content f is cached in at least one of the remote clusters. Hence, $\sum_{f=1}^{m} P_{k,f} (1 - c_{k,f})\min(\sum_{j \in \mathcal{K}\backslash\{k\}} c_{j,f}, 1)$ is the probability that the requested file f is cached in any of the remote clusters given that it is not cached in the local cluster k. The corresponding service rate is $\mu_{rc} = \frac{1}{\tau_{rc} N_a}$, where N_a represents the number of cooperating clusters simultaneously served by the remote cluster mode, i.e, the number of clusters that share the cellular rate. Finally, for the backhaul mode, the request arrival rate is written as

$$\lambda_{k,bh} = \lambda_k \sum_{f=1}^{m} P_{k,f} \prod_{k=1}^{K} (1 - c_{k,f}), \tag{10.4}$$

where $\sum_{f=1}^{m} P_{k,f} \prod_{k=1}^{K} (1 - c_{k,f})$ is the probability that the requested file f is not cached entirely in the cell, so this content could be downloaded only from the core network. The corresponding service rate is $\mu_{bh} = \frac{1}{\tau_{bh} N_b}$, where N_b is defined as the number of clusters simultaneously served via the backhaul mode.

The traffic intensity of a queue is defined as the ratio of mean service time to mean inter-arrival time. We introduce ρ_k as the metric representing the traffic intensity at cluster k, expressed as $\rho_k = \frac{\lambda_{k,lc}}{\mu_{lc}} + \frac{\lambda_{k,rc}}{\mu_{rc}} + \frac{\lambda_{k,bh}}{\mu_{bh}}$. Similar to (Collings, 1974), we consider $\rho_k < 1$ as the stability condition, otherwise, the overall delay will be infinite. The traffic intensity at any cluster is simultaneously related to the request arrival rate and the transmission rates of the three serving modes.

10.5.3.2 Network Average Delay

The mean queue size for an MPSQ with arrival rate λ (s^{-1}) is $\rho + \frac{\lambda \sum_i \lambda_i/\mu_i^2}{1-\rho}$, where ρ is the traffic intensity, and λ_i and μ_i are respectively the arrival and service rates of a service group i (Collings, 1974).

Hence, the average delay per request in cluster k is given by

$$D_k = \frac{\rho_k}{\lambda_k} + \frac{\frac{\lambda_{k \cdot lc}}{\mu_{lc}^2} + \frac{\lambda_{k \cdot rc}}{\mu_{rc}^2} + \frac{\lambda_{k \cdot bh}}{\mu_{bh}^2}}{1 - \rho_k}, \tag{10.5}$$

and hence the network weighted average delay per request is $D = \frac{1}{\lambda} \sum_{k=1}^{K} \lambda_k D_k$, where $\lambda = \sum_{i=1}^{K} \lambda_i$. We observe from (10.5) that the per request delay D_k, and correspondingly, the network average delay D, depends on the arrival rates of the three transmission modes, which are in turn functions of the content caching scheme. Because of the limited caching capacity on mobile devices, we would like to optimize the cache placement in each cluster to minimize the network weighted average delay. The delay optimization problem is then formulated as

$$\underset{c_{k \cdot f}}{\text{minimize}} \quad D \tag{10.6}$$

$$\text{subject to} \quad \sum_{f=1}^{m} c_{k \cdot f} \leq N, \tag{10.7}$$

$$c_{k \cdot f} \in \{0, 1\}, \tag{10.8}$$

where (10.7) and (10.8) are the constraints that the maximum cache size is N files per cluster, and the file is either cached entirely or is not cached at all, i.e. no partial caching is allowed. The objective function in (10.6) is not a convex function of the cache placement elements $c_{kf} \in \{0, 1\}$. Moreover, this equation can be reduced to a well known $0 - 1$ knapsack problem, which is already proven to be non-deterministic polynolmial-time hard in (Wang and Chen, 1996).

10.5.4 Proposed Caching Schemes

10.5.4.1 Caching Popular Files

In each cluster, the most popular files for the devices in the cluster are cached without repetition. Since popular files are different among clusters (but overlapped), applying caching popular files (CPF) might end up replicating the same file in many clusters (Caire and Molisch, 2015). The CPF scheme is computationally straightforward if the most popular content is known. Additionally, the CPF scheme is easy to implement in an independent manner since it is executed in a per cluster level regardless of the caching status of other clusters, which is different from the greedy algorithm proposed in the next subsection. However, it achieves high performance only if the popularity exponent β is large enough, i.e. when the content popularity distribution is skewed since a small portion of content is highly demanded that can be cached entirely in each cluster.

10.5.4.2 Greedy Caching Algorithm

In this subsection, we introduce a computationally efficient caching algorithm denoted as greedy caching algorithm (GCA). After some mathematical manipulations, we find that the constraints in (10.7) and (10.8) can be rewritten as a uniform partition matroid on a ground set that characterizes the caching elements on all clusters. Moreover, the objective function in (10.6) is a monotone non-increasing supermodular function. The greedy solution for this problem structure has been proven to be locally optimal within a factor $(1 - e^{-1}) \approx 63\%$ of the optimum (Calinescu et al., 2007; Dehghan et al., 2015; Sun et al., 2016). The GCA for the proposed D2D caching system with inter-cluster cooperation is illustrated in Algorithm 1, where S_k^f is an element denoting the placement of file f into the VCC of cluster k. The greedy algorithm works as follows. We first define the attributes of the system in the first line of the algorithm's pseudocode. We then initialize the cache memory of all clusters to zero. We set the number of iterations to be NK, which means that at each iteration, we cache one file in one cluster, resulting in caching N different files in K clusters after NK iterations. In each iteration, all combinations of caching a file $f \in \mathcal{F}$ in a cluster $k \in \mathcal{K}$ are tried, and the network service delay is calculated. A file f^* is chosen to be cached in the kth cluster, which achieves the highest reduction in the network service delay.

The greedy algorithm is run at the BS level, and the BS then instructs the clusters' devices to cache the files according to the output of this algorithm. The deterministic caching approach (both CPF and GCA) can only be realized if the devices stay at the same locations for several hours. Otherwise, the performance obtained with the deterministic caching strategy serves as a useful upper bound for more realistic schemes (Caire and Molisch, 2015). As examples of the greedy algorithm, the authors in (Dehghan et al.,

---

Algorithm 1. Greedy caching algorithm.

Input: $K, m, N, \beta, \overline{S}, R_D, \overline{R_{WL}}, \overline{R_{BH}}$;

1 **Initialization:** $C \leftarrow (0)_{K \times F}$;

2 /* Check if all clusters (devices' memories in each cluster) are fully cached. */

3 **while** $\sum_{k=1}^{K} \sum_{f=1}^{m} c_{k,f} < NK$ **do**

4 $\quad$ $(k^*, f^*) \leftarrow \text{argmax}_{(k,f)} D(C) - D(C \cup S_k^f)$;

5 $\quad$ /* File achieving highest marginal value is cached. */

6 $\quad$ $c_{k^* f^*} = 1$;

7 **end**

$\quad$ **Output:** Cache placement C;

2015) showed that the problem of optimal joint caching and routing can be formulated as maximization of a monotone sub-modular function subject to matroid constraints, and hence can be solved by the greedy algorithm.

Next, we define the outage probability for our proposed cooperative D2D caching system, and compare it with a clustered D2D caching system without inter-cluster cooperation (Ji et al., 2015).

10.5.4.3 Outage Probability

For a reference clustered D2D caching network without inter-cluster cooperation (Ji et al., 2015), the probability of no outage is defined as the probability that a randomly chosen user u can download a requested file from nearby devices in the same cluster (Ji et al., 2015). Conversely, a user u is said to be in outage when its requested file is not cached within the allowed transmission range (i.e. not cached in a neighbour user in the same cluster). In our cooperative clustered model, a user u is said to be in outage when the requested file is neither stored in the local cluster nor any of the remote clusters. We denote this outage probability as p_o, which also represents the percentage of devices who are in outage in relation to the total number of devices; the probability of no outage is then denoted as $1 - p_o$. We obtain the no outage probability as a function of the system parameters as below:

$$1 - p_o \approx \frac{\frac{1}{1-\beta}(My+1)^{1-\beta} - \frac{1}{1-\beta}}{\frac{1}{1-\beta}(m+1)^{1-\beta} - \frac{1}{1-\beta}} + \frac{1}{\frac{1}{1-\beta}(m+1)^{1-\beta} - \frac{1}{1-\beta}} \times$$

$$\sum_{j=2}^{\frac{n}{y}} \left(\frac{\left(\frac{j-1}{j}m_0 + My + 1\right)^{1-\beta} - (c')^{1-\beta}}{1-\beta} - \frac{(c')^{\beta} + \left(\frac{j-1}{j}m_0 + My + 1\right)^{\beta}}{2} \right),$$

$$= \overline{P}_{0,nc} + \overline{P}_{0,wc} \tag{10.9}$$

where $c' = \max\left(\frac{j-1}{j}m_0, \frac{j-2}{j-1}m_0 + My\right)$, and $\overline{P}_{0,nc}, \overline{P}_{0,wc}$ represent respectively the probability of no outage for a non-cooperative system and the improvement (increase) in the probability of no outage due to the inter-cluster cooperation. In Figure 10.6, we plot the outage probability of our proposed system with inter-cluster cooperation compared to a reference system without inter-cluster cooperation. We note that as the number of devices per cluster increases, the outage probability correspondingly decreases. This is attributed to the fact that the probability of obtaining the requested files from the local cluster increases with the number of devices per cluster.

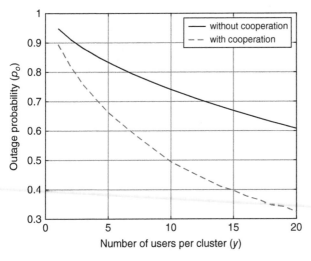

Figure 10.6 Outage probability of a D2D clustered caching system with cooperation compared to a reference system without cooperation (Ji et al., 2015) ($m = 108, n = 120, M = 1, m_0 = 60, \beta = 0.5$).

10.6   Results and Discussions

Next, we evaluate the performance of our proposed inter-cluster cooperative architecture using simulation and analytical results. Results are obtained with the following parameters: $\lambda_k = 0.5$ requests/s, $m_0 = 60$ files, $m = 108$ files, $\overline{S} = 4$ Mbits, $K = 5$ clusters, $n = 25$ devices, $M = 4$ files, and $N = 20$ files. $R_{WL} = 50$ Mbps and $R_{BH} = 5$ Mbps as in (Sun et al., 2016). For a typical D2D communication system with transmission power of 20 dBm, transmission range of 10 m, and free space path loss model as in (Ji et al., 2016a), we have $R_D = 120$ Mbps.

In Figure 10.7, the theoretical and simulated results for the network average delay under the CPF scheme are plotted together, and they are consistent. We see that the network average delay is significantly improved by increasing the cluster cache size N. Moreover, as β increases, the average delay decreases. This is attributed to the fact that a small portion of content forms most of the requests that can be cached locally in each cluster and delivered via high data rate D2D communication.

In the following, we evaluate and compare the performance of various caching schemes. In Figure 10.8, our proposed inter-cluster

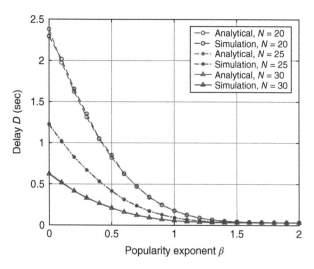

Figure 10.7   Network average delay versus popularity exponent β, under the caching popular files scheme.

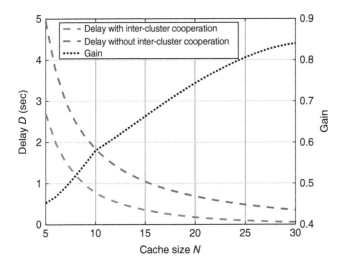

Figure 10.8 Network average delay (left hand side *y*-axis) and delay reduction gain (right hand side *y*-axis) versus cluster cache size *N*, under the caching popular files scheme.

cooperative caching system is compared with a D2D caching system without cooperation under the CPF scheme. For a D2D caching system without cooperation, requests for files that are not cached in the local cluster are downloaded directly from the core network. For the sake of concise comparison, we define the delay reduction gain as

$$\text{Gain} = 1 - \frac{D_c}{D_{nc}}, \tag{10.10}$$

where D_c and D_{nc} represent the delay with and without inter-cluster cooperation, respectively. Figure 10.8 shows that, for a small cluster cache size, the delay reduction (gain) of our proposed inter-cluster cooperative caching is higher than 45% with respect to a D2D caching system without inter-cluster cooperation and greater than 80% if the cluster cache size is large.

To show the energy-delay reduction gain trade-off among the devices, in Figure 10.9, we plot the per cluster energy consumption during the local and remote cluster modes and the gain attained from inter-cluster cooperation against the cluster cache size *N*. $P_{lc} = 20$ dBm and $P_{rc} = 23$ dBm denote respectively the transmission power in the local cluster and remote cluster modes. In each transmission

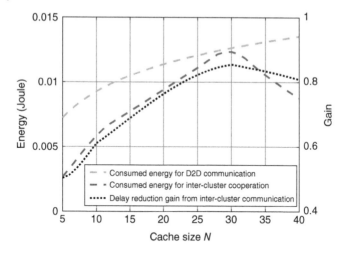

Figure 10.9 Energy consumption per cluster during the local and remote cluster transmissions (left hand side y-axis) and the delay reduction gain attained from inter-cluster cooperation (right hand side y-axis) versus cluster cache size N.

mode, the energy per request is the transmission power times the transmission duration. The transmission duration is given by the ratio of file size over the transmission rate. We see that the consumed energy during the local cluster transmission, i.e. D2D communication, monotonically increases with the cluster cache size N. With the increasing of N, more requests are served via the local cluster mode M_{lc}. For the consumed energy during the remote cluster transmission, we see that it initially increases with N, then it decreases, and the same behaviour is observed for the delay-reduction gain. This can be interpreted as follows. When N increases, the number of requests served from the remote clusters increases since the remote clusters' VCCs increase. When N becomes much larger, the local cluster cache becomes sufficiently large to serve most of the requests, as opposed to being served by the remote cluster mode.

For comparison purposes, Figure 10.10 shows the average delay for the proposed caching schemes and random caching (RC) against various system parameters. Figure 10.10(a) shows the network average delay plotted against the request arrival rate λ_k for three content placement techniques, namely, GCA, CPF, and RC. In RC, content stored in clusters are randomly chosen from the file library. The most popular

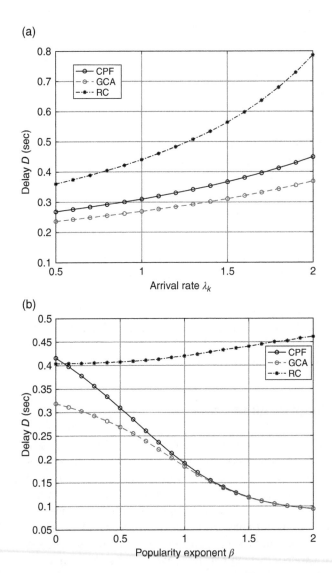

Figure 10.10 Evaluation and comparison of the average delay for the proposed caching schemes and random caching for various system parameters ($R_D = 50$ Mbps, $\overline{R_{WL}} = 15$ Mbps, $\overline{R_{BH}} = 10$ Mbps, $N = 20$, $\beta = 0.5$ for (a) and $\lambda_k = 0.5$ requests/s for (b)).

files are cached in the CPF scheme, and the GCA works as illustrated in Algorithm 1. We see that the average delay for all content caching strategies increases with λ_k since a larger request rate increases the probability of a longer waiting time for each request. It is also observed that the GCA, which is locally optimal, achieves significant performance gains over the CPF and RC solutions. Figure 10.10(b) shows that the GCA is superior to the CPF only for small values of the popularity exponent β. If the popularity exponent β is high enough, CPF and GCA will achieve the same performance. When β increases, the CDF of the Zipf distribution becomes more skewed. This implies that only a smaller portion of the files is highly demanded by the devices. The lower the number of files requested by the devices, the higher the probability of having such files cached in the clusters' VCCs. If all these files are cached locally in each cluster, the global minimum solution for the delay minimization problem is attained. This interpretation explains why when β increases, the CPF and GCA solutions converge to the global optimal solution. We also note that the CPF and RC schemes roughly achieve the same delay when $\beta = 0$. This stems from the fact that, with $\beta = 0$, all files have equal popularity, and correspondingly, CPF is equivalent to RC. Moreover, RC fails to reduce the delay as β increases, since caching files at random results in a low probability of serving the requested files from local clusters.

10.7   Chapter Summary

In this chapter, we demonstrated the role of caching towards a low latency wireless network. Among various resources of wireless networks, e.g., time, frequency and space, caching capacity stood up as a new type of resource that can effectively reduce network delay. We explained that when caches are fully utilized, the need for high-capacity backhaul links can be greatly reduced and the download delay can be minimized accordingly. We first discussed the key features of wireless caching networks. We then discussed different architectures and classifications of wireless caching networks. We followed up by describing the role of wireless caching in delivering low latency communication. We studied a cellular network consisting of one SBS and a set of devices. The cell is divided into a set of equally sized virtual clusters, where the devices in the same cluster exchange cache content via D2D communication, while the devices in different clusters cooperate by exchanging their cache content via cellular transmission. We formulated the delay minimization problem in terms of the content cache placement. Due to the problem complexity, we proposed two content caching policies, namely, CPF and GCA. Finally, we showed that the proposed GCA attains a local optimal solution within a factor $(1 - e^{-1}) \approx 63\%$ of the global optimum.

Bibliography

Hasti Ahlehagh and Sujit Dey. Hierarchical video caching in wireless cloud: Approaches and algorithms. In *IEEE International Conference on Communications (ICC)*, pages 7082–7087, June 2012.

R. Amer, M. M. Butt, M. Bennis, and N. Marchetti. Inter-cluster cooperation for wireless D2D caching networks. *IEEE Transactions on Wireless Communications*, 17(9):6108–6121, Sep. 2018a.

Ramy Amer, M. Majid Butt, Hesham ElSawy, Mehdi Bennis, Jacek Kibiłda, and Nicola Marchetti. On minimizing energy consumption for D2D clustered caching networks. In *proc. of the IEEE Global Communications Conference (GLOBECOM)*, Abu Dhabi, UAE, Dec. 2018b.

Ramy Amer, Hesham Elsawy, M. Majid Butt, Eduard A Jorswieck, Mehdi Bennis, and Nicola Marchetti. Optimizing joint probabilistic caching and communication for clustered D2D networks. *arXiv preprint arXiv:1810.05510*, 2018c.

Ramy Amer, Hesham ElSawy, Jacek Kibiłda, M. Majid Butt, and Nicola Marchetti. Cooperative transmission and probabilistic caching for clustered D2D networks. In *proc. of the IEEE Wireless Communications and Networking Conference (WCNC)*, Marrakech, Morocco, April. 2019.

J. G. Andrews. Seven ways that hetnets are a cellular paradigm shift. *IEEE Communications Magazine*, 51(3):136–144, March 2013.

Jeffrey G Andrews, Stefano Buzzi, Wan Choi, Stephen V Hanly, Angel Lozano, Anthony CK Soong, and Jianzhong Charlie Zhang. What will 5G be? *IEEE Journal on selected areas in communications*, 32 (6): 1065–1082, 2014.

B. Azari, O. Simeone, U. Spagnolini, and A. M. Tulino. Hypergraph-based analysis of clustered co-operative beamforming with application to edge caching. *IEEE Wireless Communications Letters*, 5(1):84–87, Feb 2016.

Ejder Baştu, Mehdi Bennis, Marios Kountouris, and Mérouane Debbah. Cache-enabled small cell networks: Modeling and tradeoffs. *EURASIP Journal on Wireless Communications and Networking*, 2015(1):41, 2015.

E. Bastug, M. Bennis, and M. Debbah. Living on the edge: The role of proactive caching in 5G wireless networks. *IEEE Communications Magazine*, 52(8):82–89, Aug 2014.

Ejder Bastug, Jean-Louis Guénégo, and Mérouane Debbah. Proactive small cell networks. In *ICT 2013*, pages 1–5, May 2013.

Mohamed Baza, Mahmoud Nabil, Muhammad Ismail, Mohamed Mahmoud, Erchin Serpedin, and Mohammad Rahman. Blockchain-based charging coordination mechanism for smart grid energy storage units. *arXiv preprint arXiv:1811.02001*, 2018.

Mohamed Baza, Mahmoud Nabil, Noureddine Lasla, Kemal Fidan, Mohamed Mahmoud, and Mohamed Abdallah. Blockchain-based firmware update scheme tailored for autonomous vehicles. In *proc. of the IEEE Wireless Communications and Networking Conference (WCNC)*, Marrakech, Morocco, April. 2019.

Niclas Bewermeier Kemal Fidan Mohamed Mahmoud Mohamed Abdallah Baza, Mahmoud Nabil. Detecting sybil attacks using proofs of work and location in VANETs. *arXiv preprint arXiv:1904.05845*, 2019.

Mehdi Bennis, Mérouane Debbah, and H Vincent Poor. Ultra-reliable and low-latency wireless communication: Tail, risk and scale. *arXiv preprint arXiv:1801.01270*, 2018.

S. Bi, R. Zhang, Z. Ding, and S. Cui. Wireless communications in the era of big data. *IEEE Communications Magazine*, 53(10):190–199, October 2015.

B. Blaszczyszyn and A. Giovanidis. Optimal geographic caching in cellular networks. In *IEEE International Conference on Communications (ICC)*, pages 3358–3363, June 2015.

L. Breslau, Pei Cao, Li Fan, G. Phillips, and S. Shenker. Web caching and zipf-like distributions: evidence and implications. In *IEEE Conference on Computer Communications (INFOCOM)*, NY, USA, March. 1999.

Giuseppe Caire and Andreas F Molisch. Femtocaching and D2D communications: A new paradigm for video-aware wireless networks. *Intel Technology Journal*, 19(1), 2015.

G. Calinescu, C. Chekuri, M. Pal, and J. Vondrak. Maximizing a supermodular set function subject to a matroid constraint. In *12th International IPCO Conference*, NY, USA, June. 2007.

A. Checko, H. L. Christiansen, Y. Yan, L. Scolari, G. Kardaras, M. S. Berger, and L. Dittmann. Cloud ran for mobile networks? a technology overview. *IEEE Communications Surveys Tutorials*, 17(1): 405–426, Firstquarter 2015.

B. Chen, C. Yang, and G. Wang. High-throughput opportunistic cooperative device-to-device communications with caching. *IEEE Transactions on Vehicular Technology*, 66(8):7527–7539, Aug 2017.

M. Chiang and T. Zhang. Fog and iot: An overview of research opportunities. *IEEE Internet of Things Journal*, 3(6):854–864, Dec 2016.

Visual Networking Index Cisco. Global mobile data traffic forecast update, 2016–2021 white paper. *Document ID, 1454457600805266*, 2016.

T. W. R. Collings. A queueing problem in which customers have different service distributions. *Appl. Statist.*, 34(1):75–82, 1974.

Mostafa Dehghan, Anand Seetharam, Bo Jiang, Ting He, Theodoros Salonidis, Jim Kurose, Don Towsley, and Ramesh Sitaraman. On the complexity of optimal routing and content caching in heterogeneous networks. In *IEEE Conference on Computer Communications (INFOCOM)*, Hong Kong, April. 2015.

Vincent Etter, Mohamed Kafsi, and Ehsan Kazemi. Been there, done that: What your mobility traces reveal about your behavior. In *Mobile Data Challenge by Nokia Workshop, in conjunction with Int. Conf. on Pervasive Computing*, number EPFL-CONF-178426, 2012.

N. Golrezaei, K. Shanmugam, A. G. Dimakis, A. F. Molisch, and G. Caire. Femtocaching: Wireless video content delivery through distributed caching helpers. In *2012 Proceedings IEEE INFOCOM*, pages 1107–1115, March 2012.

N. Golrezaei, A. F. Molisch, A. G. Dimakis, and G. Caire. Femtocaching and device-to-device collaboration: A new architecture for wireless video distribution. *IEEE Communications Magazine*, 51(4):142–149, April 2013.

N. Golrezaei, P. Mansourifard, A. F. Molisch, and A. G. Dimakis. Base-station assisted device-to-device communications for high-throughput wireless video networks. *IEEE Transactions on Wireless Communications*, 13(7):3665–3676, July 2014a.

Negin Golrezaei et al. Base-station assisted device-to-device communications for high-throughput wireless video networks. *IEEE Transactions on Wireless Communications*, 13(7):3665–3676, July 2014b.

Brett D Higgins, Jason Flinn, Thomas J Giuli, Brian Noble, Christopher Peplin, and David Watson. Informed mobile prefetching. In *Proceedings of the 10th international conference on Mobile systems, applications, and services*, pages 155–168. ACM, 2012.

S. Hosny, F. Alotaibi, H. El Gamal, and A. Eryilmaz. Towards a mobile content marketplace. In *IEEE 16th International Workshop on Signal Processing Advances in Wireless Communications (SPAWC)*, pages 675–679, June 2015.

S. W. Jeon, S. N. Hong, M. Ji, G. Caire, and A. F. Molisch. Wireless multihop device-to-device caching networks. *IEEE Transactions on Information Theory*, 63(3):1662–1676, March 2017.

M. Ji, G. Caire, and A. F. Molisch. The throughput-outage tradeoff of wireless one-hop caching networks. *IEEE Transactions on Information Theory*, 61(12):6833–6859, Dec 2015.

M. Ji, G. Caire, and A. F. Molisch. Wireless device-to-device caching networks: Basic principles and system performance. *IEEE Journal on Selected Areas in Communications*, 34(1):176–189, Jan 2016a.

M. Ji, G. Caire, and A. F. Molisch. Fundamental limits of caching in wireless D2D networks. *IEEE Transactions on Information Theory*, 62 (2):849–869, Feb 2016b.

M. K. Karray and M. Jovanovic. A queueing theoretic approach to the dimensioning of wireless cellular networks serving variable-bit-rate calls. *IEEE Transactions on Vehicular Technology*, 62(6):2713–2723, July 2013.

D. Liu and C. Yang. Energy efficiency of downlink networks with caching at base stations. *IEEE Journal on Selected Areas in Communications*, 34(4):907–922, April 2016.

D. Liu, B. Chen, C. Yang, and A. F. Molisch. Caching at the wireless edge: design aspects, challenges, and future directions. *IEEE Communications Magazine*, 54(9):22–28, Sep. 2016.

M. A. Maddah-Ali and U. Niesen. Fundamental limits of caching. *IEEE Transactions on Information Theory*, 60(5):2856–2867, May 2014.

M. A. Maddah-Ali and U. Niesen. Cache-aided interference channels. In *IEEE International Symposium on Information Theory (ISIT)*, pages 809–813, June 2015.

Mugen Peng, Shi Yan, Kecheng Zhang, and Chonggang Wang. Fog-computing-based radio access networks: issues and challenges. *Ieee Network*, 30(4):46–53, 2016.

X. Peng, J. Shen, J. Zhang, and K. B. Letaief. Backhaul-aware caching placement for wireless networks. In *IEEE Global Communications Conference (GLOBECOM)*, pages 1–6, Dec 2015.

Thomas Pfeiffer. Next generation mobile fronthaul and midhaul architectures. *Journal of Optical Communications and Networking*, 7 (11):B38–B45, 2015.

Jack Robinson, Peter Muller, Timothy Noke, Teng Lew Lim, Wallace Glausi, Larry Fullerton, and Dusan Hamar. Dynamic information management system and method for content delivery and sharing in content-, metadata- and viewer-based, live social networking among users concurrently engaged in the same and/or similar content, April 22 2014. US Patent 8,707,185.

A. Sengupta, R. Tandon, and T. C. Clancy. Improved approximation of storage-rate tradeoff for caching via new outer bounds. In *2015 IEEE International Symposium on Information Theory (ISIT)*, pages 1691–1695, June 2015.

A. Sengupta, R. Tandon, and O. Simeone. Cache aided wireless networks: Tradeoffs between storage and latency. In *Annual Conference on Information Science and Systems (CISS)*, pages 320–325, March 2016.

K. Shanmugam, N. Golrezaei, A. G. Dimakis, A. F. Molisch, and G. Caire. Femtocaching: Wireless content delivery through distributed caching helpers. *IEEE Transactions on Information Theory*, 59(12): 8402–8413, Dec 2013.

O. Simeone, A. Maeder, M. Peng, O. Sahin, and W. Yu. Cloud radio access network: Virtualizing wireless access for dense heterogeneous systems. *Journal of Communications and Networks*, 18(2):135–149, April 2016.

Y. Sun, Z. Chen, and H. Liu. Delay analysis and optimization in cache-enabled multi-cell cooperative networks. In *IEEE Global Communications Conference (GLOBECOM)*, Washington DC, USA, Dec. 2016.

R. Tandon and O. Simeone. Cloud-aided wireless networks with edge caching: Fundamental latency tradeoffs in fog radio access networks. In *2016 IEEE International Symposium on Information Theory (ISIT)*, pages 2029–2033, July 2016.

M. Tao, E. Chen, H. Zhou, and W. Yu. Content-centric sparse multicast beamforming for cache-enabled cloud RAN. *IEEE Transactions on Wireless Communications*, 15(9):6118–6131, Sep. 2016.

T. X. Tran, A. Hajisami, and D. Pompili. Cooperative hierarchical caching in 5g cloud radio access networks. *IEEE Network*, 31(4):35–41, July 2017.

Intelligent Small Cell Trial. Rethinking the small cell business model. 2012.

Chihping Wang and Ming-Syan Chen. On the complexity of distributed query optimization. *IEEE Transactions on Knowledge and Data Engineering*, 8(4):650–662, Aug 1996.

11

Application of Terahertz Sensing at Nano-Scale for Precision Agriculture

Adnan Zahid[1], Hasan T. Abbas[1], Aifeng Ren[1], Akram Alomainy[2], Muhammad A. Imran[1], and Qammer H. Abbasi[1]

[1] *James Watt School of Engineering, University of Glasgow, UK*
[2] *School of Electronic Engineering and Computer Science, Queen Mary University of London, UK*

11.1 Introduction

With the rising scarcity of water resources in the plant science sector, many significant and modern techniques have been evolved in applied plant biology at various level over the past few decades. In most countries, agriculture is considered as the spine in the overall development of countries due to its significant role in enhancing the economic development of the country Beyrouthya (2014). To meet with the challenging growth of population, it is important to have smart, sustainable, and precision agriculture to increase the productions of crops as well as providing concrete information of tailoring soils and crops by employing various sensors into the agricultural field. Precision agriculture includes the usage of fertilizers, pesticides, and herbicides to minimum degree for maximum production Beyrouthya (2014).

The sensors that have been functional so far in terms of achieving accuracy in agriculture are temperature sensors, humidity sensors, soil moisture sensors, pH sensors, light sensors, colour sensors, and bio sensors (used for detection of nutrients contents in soil) Mukhopadhyay (2014); Ravichandran (2010a). Various systems have been proposed where these sensors have been utilised at large scale and have produced the optimum results by mitigating the overall

Wireless Automation as an Enabler for the Next Industrial Revolution, First Edition.
Edited by Muhammad A. Imran, Sajjad Hussain and Qammer H. Abbasi.
© 2020 John Wiley & Sons Ltd. Published 2020 by John Wiley & Sons Ltd.

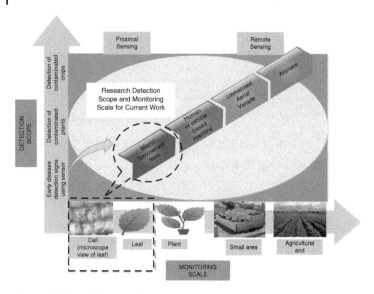

Figure 11.1 Various sensing technologies applied at different stages for the detection of the contaminated crops.

workload of farmers in cultivation and address the drawbacks of conventional farming, as shown in Figure 11.1. There have been occasions where multiple sensors were brought into service to determine every parameter that affected the overall growth of crops, and web based tailored applications offered flexibility to monitor the agricultural land from any location. There is now a need to develop a more innovative and novel approach due to the rapid growth in the population, lack of crop productivity in existing technologies, unnecessary and untimely use of fertilizers and pesticides inviting health and purification expenses and incurring nutrients and toxins increments, and failure in the detection of disease and bacteria in crops at an early stage Prasad et al. (2017).

Nanotechnology has huge potential to tackle the above mentioned concerns and can be used as a useful means to meet the needs of a growing population, increase in the productivity of crops, and the detection of pathogens (bacteria, virus) at the very initial level. Nanotechnology basically employs nanomaterials, which have emerged as a huge benefit with massive prospects in the agriculture sector. Development of nanosensors can be utilized for sensing a wide variety of fertilizers, herbicides, pesticides, insecticide, pathogens, etc. toward the support of sustainable agriculture for increasing crop productivity. An innovative approach for rapid disease detection in plants, especially dealing with crop pathogens, improving the ability of plants to absorb nutrients and decreasing the practise of using conventional pesticides and herbicides, has verified that nanotechnology has enormous capability to transform the agriculture sector Mukhopadhyay (2014); Prasad et al. (2017).

11.1.1 Limitations of Conventional Methods

Considering the conventional approaches, standard sensors, as mentioned earlier, have been employed to meet the huge requirement of crop productivity and appropriate usage of fertilizers, and are capable of detecting small amounts of impurities in soil and pathogens in plants, nutrients deficiencies in plants. Regrettably, they have not obtained prolific results in agriculture sectors and clearly appear to be unfeasible and do not achieve the required results. Over the past few decades Omanović-Mikličanina and Maksimović (2016); Ravichandran (2010b), development in science and technology, especially in the agriculture sector, has not been able to offer longer lasting and reliable solutions to sustain farm productivity. In spite of employing conventional approaches such as macrosensors, the required productivity outcome, especially in developing countries, has not been achieved. In addition, it was also far difficult to reduce the environmental and other resource expenses related to agricultural production by lowering the nutrient losses in fertilization and the amount of spread chemicals Gruère (2012) Omanović-Mikličanina and Maksimović (2016)

11.1.2 Transformation from Micro- to Nanotechnology

Over the past several decades, developments in science and technology especially in the agricultural field, have not been able to offer more long lasting and reliable solutions to sustain farm productivity Duhan et al. (2017). Despite employing conventional approaches such as macrosensors, could not have brought required productivity outcome especially in developing countries. In addition, it was also far difficult to reduce the environmental and other resource expenses related to agricultural production by lower the nutrient losses in fertilization and the amount of spread chemicals Gruère (2012).

The practical implementation of microsensors in the field of agriculture have grown immensely over the past few years due to the strong advantages over available conventional sensors (macro approach) in the market. Microsensors, which are purely constructed based on semiconductor technology compared to conventional sensors, yield supplementary advantages such as they are smaller in size, lighter, easily portable, extremely energy efficient, robust and have a quick response time . These microsensors are fabricated using microfabrication methods and, compared to available macro level sensors, they exhibit low power consumption compared to standard sensors Gruère (2012)

Since this new nanotechnology came into existence, the development and sustainable growth of agriculture has depended on this new and innovative technique Alfadul et al. (2017). Compared to the micro approach, this technology is deemed to have more reliability, fast response and is a low-cost system for the overall monitoring and maintaining the health of the leaves Prasad et al. (2017). For instance, new nanowire based materials used in nanotechnology contains exceptional detection and sensing properties and it can deliver advanced sensitivity, better selectivity, and probably provides better stability at lower cost Prasad et al. (2017).

In modern agriculture, the progress of biosensors plays a vital role in improving both the quantity and quality of yields due to the properties of nanomaterials. Furthermore, researchers have effectively and sensibly used nanomaterials and have managed to miniaturize biosensors to more tiny and compact devices such as nanosensors to monitor disease detection in leaves at a molecular level and are very capable of real-time analysis. In addition, the enormous progression in nanobiosensors is because of the huge scientific demand for quick, sensitive and commercially worthwhile nano-biosensor systems in important spheres such as healthcare, agriculture, and environmental monitoring. Presently, the progress of developing nanotechnology based nano-biosensors is at its infancy stage. Regarding the utilization of nanomaterials for the fabrication of nano-biosensors, metals (gold, silver, cobalt etc.) have been widely investigated for their useful applications and purpose in biosensors Mukhopadhyay (2014).

11.1.3 Evolution of Nanotechnology

Since the evolution of nanotechnology, an opportunity to dig more deeply into molecular level has been made possible, such as in cells, leaves, and plants, to create the ability to detect disease at a very early stage employing nanosensors made up of nanomaterials. This nanotechnology can play a vital role in overall productivity by controlling the nutrients, observing the water quality and any presence of pesticides for effective and sustainable progress in the agricultural sector Gruère (2012); Prasad et al. (2014). Compared to the micro approach, this technology is deemed to have more reliability, fast response, and is a low-cost system for overall monitoring and maintaining the health of leaves Prasad et al. (2014); Setter et al. (2006). For instance, new nanowire based materials used in nanotechnology contain exceptional detecting and sensing properties and they can deliver advanced sensitivity, better selectivity, and probably provides better stability at lower cost. Prasad et al. (2017)

In this regard, nanosensor applications operating at terahertz frequencies Afsharinejad et al. (2016) introduce distinctive opportunities for sensing plant responses to changes in environmental conditions, contamination detection, pathogen attacks and other stresses. It also has the capability to monitor plant signalling and responses during the process of growth and development, water content level in plant leaves and bacteria detection in crops at the molecular level Chen and Yada (2011); Lu and Bowles (2013). The evolving application of terahertz (THz) technology ? has fascinated numerous horticulturists, researchers and scientists like plant physiologists or biochemists in various fields of applied plant biology.

11.1.4 Potential Benefits of Nanotechnology in Agriculture

Presently, the research and development pipeline has the proficiency to make agriculture methods more competent and resourceful by increasing yields and product quality. In developed countries, researchers are in a position to carry out experiments for testing nanosensors and non-agricultural chemicals, nanoparticles for soil cleaning and nanopore filters, and nanoceramic devices. An increasing number of applications are expected for food and agriculture use, including nanosensors, potentially capable of detecting chemical contaminants, viruses, and bacteria. Nanodelivery systems, which could precisely deliver drugs or micronutrients at the right time and to the right part of the body, as well as nanocoating and films, nanoparticles Prasad et al. (2014).

In developing countries, researchers have made significant efforts in bringing out the potential of nanotechnology and have found that nanoscale agrichemical formulations can enhance the use efficiency and obtain a reduction in environmental losses. Researchers have also found that nanoporous materials can store water and with gradually draining it during water shortages, they could also increase production and help in saving water. Researchers have also found that implementation of nanotechnology can help in reducing the effects of fungal toxin and increases the weight of food animals.

11.1.5 Challenges in Nanotechnology

It is certain that nanotechnology has transformed the agriculture sector with its benefits and several prospective uses. However, there are some limitations that also need to be addressed and they are as follows.

11.1.5.1 Health and Environmental Impacts
The uniqueness and development of nanotechnology causes some ambiguity considering the long-standing consequences of nanoparticles on human health and on the environment through bio-accumulation of toxins in plants and animals. To overcome this constraints in toxicity and obtain some diminution in the environment consequences caused by nanoparticles, materials can be restructured using materials that are non-toxic and biodegradable Prasad et al. (2017).

11.1.5.2 High Production Costs
Due to the high exposure to this new technology, it does have high production costs and certainly requires more effort and focus in research and development to make it cost effective and also affordable, especially for developing countries. It will require making

arrangements for more collaboration between various countries that possess advanced research and applications of nanotechnology. Currently most of the research and development is taking place in a selected number of countries and knowledge is unequally distributedPrasad et al. (2017).

11.1.5.3 Risk Assessment

The existence of the variety of nanoparticles and a deficiency of data on their toxicity under different circumstances has delayed the formation of normalized risk assessment instruments. The collection of a variety of nanoparticles with similar characteristics increases the feasibility of the risk assessment and is not considered to be reliable. There is a strong need for a concise and sound international consensus approach on a workable definition Duhan et al. (2017); Mukhopadhyay (2014).

11.1.6 Evolving Applications of Terahertz (THz) Technology

The prevailing challenges of food security, sustainability and the effect of climate change have immensely engaged researchers in exploring the field of Terahertz in depth, considering it as new source of vital improvements for the agricultural sector. In recent times, there has been significant advancements in THz applications in numerous fields such as biomedical imaging, diagnostic applications, and safety and security due to its distinctive characteristics. Some of the advancements that have been achieved from these unique features are security imaging of invisible items, atmospheric studies, processing the quality control of food, high-frequency communication, non-contact imaging for protection of paintings and manuscripts, and medical imaging for non-invasive diagnostics of dental care and skin cancer. Moreover, terahertz technology has also shown distinct advancements in exploring the molecular changes of water content (WC) in leaves due to the high sensitivity and penetration features of terahertz radiation to the absorption of water by the leaves, as shown in Figure 11.2. Jornet and Akyildiz (2011); Song et al. (2018); Zahid et al. (2018).

Despite these aforementioned advantages and the enormous contributions of terahertz technology in various fields Zahid et al. (2018), researchers are of the view that its potential in the field of plant science

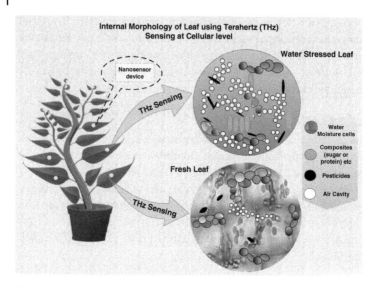

Figure 11.2 Internal morphology of fresh and water stressed leaves using terahertz sensing.

is still considered one of the least examined research areas until now Rehn et al. (2016); Zahid et al. (2018). Nevertheless, researchers have made efforts such as monitoring and controlling of environmental systems, crop productivity enhancement, protection of crops from any pathogen attacks (Born et al., 2014), and in particular, monitoring and determining the appropriate amount of water content in plant leaves by employing various methods Rehn et al. (2016). In the past, many functional techniques have been suggested, such as infrared imaging, hyper-spectral thermal imaging and magnetic spectrum imaging, to determine the comprehensive analysis of spatial and spectral information of the water content in plant leaves Saha et al. (2012); Song et al. (2018); Zahid et al. (2018).

Although there has been substantial progress achieved by utilizing these techniques, there are some limitations related to each technique and they could be enhanced further. In addition, thermal and hyper-spectral imaging are generally categorized as derivation methods, which mirror the in-depth influence triggered by the modification of water content. For instance, considering the infrared or microwave spectrum, they are either incapable of detecting minor changes in the water status in leaves or are littered with the inorganic salt content of the plant, resulting in vital disturbances in the overall measurement process. Saha et al. (2012); Song et al. (2018); Zahid et al. (2018).

In this chapter, employing the state-of-the-art terahertz technique, the focus is primarily to introduce a preliminary analysis and to investigate the WC and presence of any pesticides in leaves in the range of 0.75 to 1.2 THz frequency using the Swissto12 system (Swissto12, 2017). It is also aimed at determining the path-loss response and the complex permittivity of leaves of fresh and drought stressed leaves. Consequently, this has led to valuable information and strong correlation is identified considering the feasibility of this novel method for examining the temporal and dehydration response in plants Zahid et al. (2018).

11.1.7 Materials and Methods

11.1.7.1 Experimental Setup

Prior to starting the terahertz transmission measurements using the Swissto12 system, full calibration of two ports (wR1.0), also known as short-open-load-through, was performed using a modern calibration technique, as shown in Figure 11.3. The purpose of the whole calibration was to avoid any measurement errors produced by the system or any other external factor. As mentioned previously, the Swissto12 system has a frequency range from 0.75 to 1.2 THz, with a thickness range from 40 m to 4 mm. In addition, both reflection coefficients (S11 and S22) and transmission parameters (S12 and S21) were measured whilst performing the experiments on various leaves Zahid et al. (2018).

11.1.7.2 Sample

Six various fresh leaves were considered for the experiment namely: baby-leaf, basil, pea-shoot and spinach (Figure 11.4). All these leaves were detached from a fresh pot and placed in the laboratory under same environmental conditions for four consecutive days. These fresh plant leaves were not watered for the next four days. The tested environment temperature for the measurements was 18 °C 0.1 °C. Three different locations on the leaves were examined to observe any surface irregularities and the WC was as shown in Figure 11.5. The thickness and weight of the leaves were also measured using Vernier calipers and a digital scale, respectively, as shown in Figure 11.5.

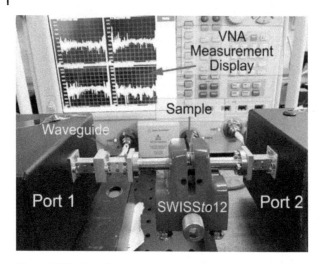

Figure 11.3 Experimental setup for measuring the transmission response by using the Swissto12 terahertz system using a frequency range from 0.75 to 1.1 THz.

11.1.7.3 Thickness of Leaves

For the examination of the measurement results of the leaves, it was significant to calculate the thickness of twelve lcaves by means of a high precision measurement tool called Vernier calipers, as shown below in Figure 11.5. Each of the leaves was tested individually for thickness under the same environment conditions. The thickness range of the leaves was found in the threshold range of the system (40 μm to 4 mm), which enabled performing experiments on them. This process was repeated three times to obtain the thickness of the leaves in different positions. During the investigation of the thickness of the leaves, it was observed that the thickness continuously increased with growing water content.

11.1.8 Measurement Results

11.1.8.1 Transmission Response

In order to determine the transmission and path-loss response of leaves, the leaves were placed in a material characterisation kit (MCK) Swissto12. It was noticed that the transmission response of the leaves was distinguishable from each other, as shown in Figure 11.6. Based on the initial measurement results, it was noticed that

Figure 11.4 Samples of different fresh leaves used for the measurement process.

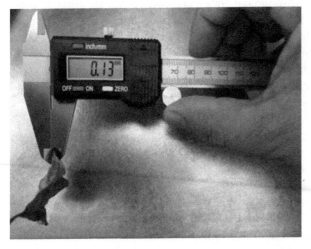

Figure 11.5 Demonstration of measuring the thickness of a leaf using Vernier calipers.

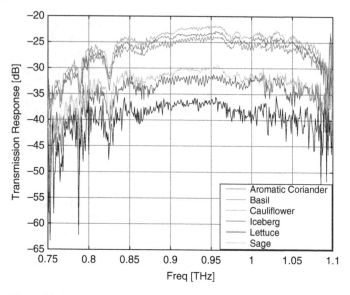

Figure 11.6 Transmission response of various leaves in the frequency range 0.75 to 1.1 THz.

different leaves exhibited various attenuation and, clearly showed correlation to the amount of water content or any pesticides presence in leaves. It also occurred that variation in transmission response could be due to different thickness of leaves, and components of plant physiology. In addition, from Figure 11.6, taking all the measurement response into consideration, it can be concluded that out of six leaves, cabbage exhibits the higher amount of water content followed by pea shoot and others.

These observations led to significant and meaningful information to study and analyse the existence of any pesticides in the leaves with terahertz frequencies. Moreover, the results also revealed that cauliflower exhibits a lower amount of water content or the presence of any pesticides in leaves as compared to other leaves. However, it also shows a sensitive region in the frequency range from 0.75 to 0.8 THz. This significant and meaningful information can be very useful to determine the volumetric amount of water content in leaves and take proactive and timely action to water the leaves to avoid any scarcity in leaves.

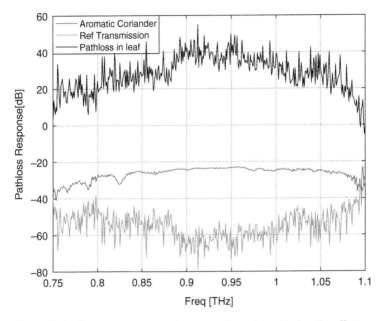

Figure 11.7 Path-loss response of an aromatic leaf considering the effect of different thickness of leaves and changes in the water content of leaves.

11.1.8.2 Path-loss Response of Leaves

Furthermore, a path-loss response was also obtained as shown in Figures 11.7, 11.8 and 11.9, proven to be more intriguing results considering the response of various leaves. It is depicted that in a certain frequency range between 0.79 Hz to 0.83 THz, as shown in Figure 11.6, a more sensitive frequency zone and precise identification of changes in water content of leaves is revealed. It was perceived that the path-loss response could be due to the distinguished surface roughness of leaves, the presence of water content in leaves and the propagation properties of leaves in the process of their growth. Comparing the results of the three leaves in Figure 11.7 it is interesting to notice that the aromatic leaf exhibit a high path-loss response compare to other leaves, i.e. sage and basil. From the figure, it can also be noticed that the least path-loss response obtained in basil compare to sage due to having some high dip in the frequency range from 1 Hz to 1.1THz.

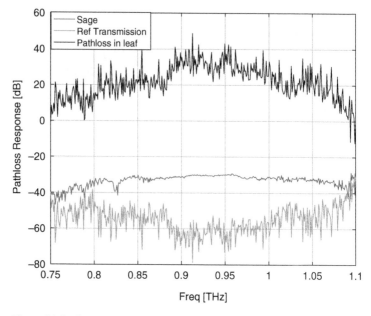

Figure 11.8 Path-loss response of a sage leaf considering the effect of different thickness of leaves and changes in the water content of leaves.

Eventually, it is revealed that there is relatively high dehydration rate or less water content in the aromatic leaf compared to other results. In addition, the results also clearly reveal the relationship between the thickness and water content of leaves. The path-loss response of leaves exhibits a strong correlation with water content and with the physical statistics of the leaves. Hence, it also highlights the information about absorption loss and reveals how much water content is absorbed by the leaves and how much is reflected back. Absorption loss can be due to surface irregularities or due to the high absorption of water content at the time of measurement.

11.1.9 Conclusion

This chapter describes various technologies that are currently being used to enhance agricultural production to meet the demands of a growing population. Due to the growing population and considering the current agricultural situation, it has been analysed that previously

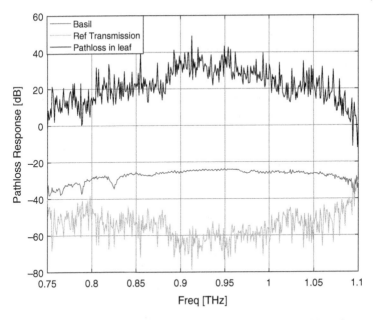

Figure 11.9 Path-loss response of a basil leaf considering the effect of different thickness of leaves and changes in the water content of leaves.

conventional technologies appear to be unfeasible and unable to produce the optimum results in terms of crop productivity, timely use of fertilizers and detection of bacteria at an early stage to mitigate the overall cost. Therefore, in order to obtain higher accuracy and an effective use of water in the agriculture sector, terahertz sensing has potential and is deemed to have a faster, reliable response for the overall monitoring and maintaining the health of the leaves. This chapter emphasizes the advancement and development of terahertz technology applications in the agriculture sector and presents some of the preliminary studies done in this domain, which has huge potential to provide proactive intervention and change the paradigm of agritech.

Bibliography

Armita Afsharinejad, Alan Davy, Brendan Jennings, and Conor Brennan. Performance analysis of plant monitoring nanosensor networks at thz frequencies. *IEEE Internet of Things Journal*, 3(1):59–69, 2016.

S Alfadul, O Altahir, and M Khan. Application of nanotechnology in the field of food production. *Acad J Sci Res*, 5(7):143–154, 2017.

Marc El Beyrouthya. Nanotechnologies: Novel solutions for sustainable agriculture. *Advances in Crop Science and Technology*, 02(03), 2014. doi: 10.4172/2329-8863.1000e118.

Hongda Chen and Rickey Yada. Nanotechnologies in agriculture: new tools for sustainable development. *Trends in Food Science & Technology*, 22(11):585–594, 2011.

Joginder Singh Duhan, Ravinder Kumar, Naresh Kumar, Pawan Kaur, Kiran Nehra, and Surekha Duhan. Nanotechnology: The new perspective in precision agriculture. *Biotechnology Reports*, 15:11–23, 2017.

Guillaume P. Gruère. Implications of nanotechnology growth in food and agriculture in OECD countries. *Food Policy*, 37(2):191–198, apr 2012. doi: 10.1016/j.foodpol.2012.01.001.

Guillaume P Gruère. Implications of nanotechnology growth in food and agriculture in oecd countries. *Food Policy*, 37(2):191–198, 2012.

J. M. Jornet and I. F. Akyildiz. Channel modeling and capacity analysis for electromagnetic wireless nanonetworks in the terahertz band. *IEEE Transactions on Wireless Communications*, 10(10):3211–3221, October 2011. ISSN 1536-1276. doi: 10.1109/TWC.2011.081011.100545.

Jianjun Lu and Marcus Bowles. How will nanotechnology affect agricultural supply chains? *International Food and Agribusiness Management Review*, 16(1030-2016-82815), 2013.

Siddhartha S. Mukhopadhyay. Nanotechnology in agriculture: prospects and constraints. *Nanotechnology, Science and Applications*, 2014: 63–71, aug 2014. doi: 10.2147/nsa.s39409.

Enisa Omanović-Mikličanina and Mirjana Maksimović. Nanosensors applications in agriculture and food industry. *Bull Chem Technol Bosnia Herzegovina*, 47:59–70, 2016.

Ram Prasad, Vivek Kumar, and Kumar Suranjit Prasad. Nanotechnology in sustainable agriculture: present concerns and future aspects. *African Journal of Biotechnology*, 13(6):705–713, 2014.

Ram Prasad, Atanu Bhattacharyya, and Quang D. Nguyen. Nanotechnology in sustainable agriculture: Recent developments, challenges, and perspectives. *Frontiers in Microbiology*, 8, June 2017. doi: 10.3389/fmicb.2017.01014.

R. Ravichandran. Nanotechnology applications in food and food processing: Innovative green approaches, opportunities and uncertainties for global market. *International Journal of Green Nanotechnology: Physics and Chemistry*, 1(2):P72–P96, may 2010a. doi: 10.1080/19430871003684440.

R Ravichandran. Nanotechnology applications in food and food processing: innovative green approaches, opportunities and uncertainties for global market. *International Journal of Green Nanotechnology: Physics and Chemistry*, 1(2):72–96, 2010b.

A. Rehn, R. Gente, T. Probst, J. C. Balzer, and M. Koch. Plant water status monitoring with THz qtds. In *Proc. German Microwave Conf. (GeMiC)*, pages 4–6, March 2016. doi: 10.1109/GEMIC.2016.7461541.

S. C. Saha, J. P. Grant, Y. Ma, A. Khalid, F. Hong, and D. R. S. Cumming. Terahertz frequency-domain spectroscopy method for vector characterization of liquid using an artificial dielectric. *IEEE Transactions on Terahertz Science and Technology*, 2(1):113–122, January 2012. ISSN 2156-342X. doi: 10.1109/TTHZ.2011.2177172.

JR Setter, Peter J Hesketh, and Gary W Hunter. Sensors: engineering structures and materials from micro to nano. *Interface*, 15(1):66–69, 2006.

Z. Song, S. Yan, Z. Zang, Y. Fu, D. Wei, H. Cui, and P. Lai. Temporal and spatial variability of water status in plant leaves by terahertz imaging. *IEEE Transactions on Terahertz Science and Technology*, 8(5):520–527, September 2018. ISSN 2156-342X. doi: 10.1109/TTHZ.2018.2851922.

A. Zahid, K. Yang, H. Heidari, C. Li, M. A. Imran, A. Alomainy, and Q. H. Abbasi. Terahertz characterisation of living plant leaves for quality of life assessment applications. In *Proc. Baltic URSI Symp. (URSI)*, pages 117–120, May 2018. doi: 10.23919/URSI.2018.8406770.

Index

Wireless Automation as an Enabler for the Next Industrial Revolution, First Edition.
Edited by Muhammad A. Imran, Sajjad Hussain and Qammer H. Abbasi.
© 2020 John Wiley & Sons Ltd. Published 2020 by John Wiley & Sons Ltd.